William J. Linton

The Ferns of the English Lake Country

With a List of Varieties

William J. Linton

The Ferns of the English Lake Country
With a List of Varieties

ISBN/EAN: 9783337227074

Printed in Europe, USA, Canada, Australia, Japan

Cover: Foto ©berggeist007 / pixelio.de

More available books at **www.hansebooks.com**

THE ROYAL FERN.

THE FERNS

OF THE

ENGLISH LAKE COUNTRY:

WITH A LIST OF VARIETIES.

BY

W. J. LINTON.

LONDON: HAMILTON, ADAMS AND CO.

WINDERMERE: J. GARNETT.

1865.

PREFACE.

FOR the general scientific part of this book I am indebted to the Editor of *Nature-printed Ferns* — Mr. Thomas Moore — whose various works upon British Ferns have exhausted nearly all that can be said upon the subject. On this ground my book could hardly be more than a mere abridgement and paraphrase of his writing. For new and original matter I have to thank Mr. F. Clowes, of Windermere, who very kindly furnished me with the lists of local varieties, at the same time according me so much help and supervision as may, I trust, be sufficient guarantee for correctness. My obligations ought also to be expressed to Mr. Barnes and other collectors, from whom, through Mr. Clowes, I obtained information for these lists. The General Description and the Chapter on Meanings of Names and Terms have been added not only to give a more universal interest to the book, but as almost necessary for the collector or student, since nearly all the genera of British Ferns (sixteen out of nineteen) are found in the Lake Country. The engravings are from nature, and as many have been given as are wanted to show the characters of genera and species, some few only having been omitted as not distinguishable in figures on so small a scale. The Remote Buckler Fern, introduced to the British Flora by Mr. Clowes, and first described in the Addenda to Mr. Moore's 8vo. edition of *Nature-printed British Ferns*, 1862, has, I believe, been engraved only once before now.

So much of Preface seemed right to explain the nature of the work and to render fitting acknowledgements and thanks. I have only farther to say that I shall be obliged to any one who will favour me by either pointing out possible errors or giving me information as to new varieties or habitats. And I ought to add that Fern-collectors may obtain good specimens from Mr. Grier, nurseryman, Ambleside, or Mr. Wood, fern-seller, Bowness.

W. J. LINTON.

BRANTWOOD, CONISTON,
    SEPT., 1865.

CONTENTS.

FERNS

OF THE ENGLISH LAKE COUNTRY.

GENERAL DESCRIPTION.

In the broad primary division of the great Vegetable Kingdom into FLOWERING and FLOWERLESS Plants FERNS are placed at the head of the second class. Possessing a distinct stem and leaves (the latter usually named fronds, to distinguish them from the leaves of other plants), *they are without flowers, ordinarily so called*, and consequently cannot produce seeds in the ordinary flower-manner. They have also a special structure: for while the Flowering Plants are either EXOGENS — plants whose stems consist of pith, wood, and bark, growing in concentric circles, and whose leaves have veins branching like net-work, — or ENDOGENS — plants without distinction of pith or bark,

B

whose stems are merely confused pithy matter or woody fibrous thread-like bundles, and whose leaf-veins are parallel, — the Flowerless Plants, called ACROGENS, *have their wood disposed in a zigzag manner, and their leaves are either without veins, or with veins of the most simple character,* scarcely branching at all, or branching only in repeated forks. The greater outer distinction, however, is that of the absence of any apparent flowers, and the arrangement of the seed (or *spores*) in seed-vessels (or *spore-cases*) upon the leaves themselves. This is speaking of the Ferns only, for other Acrogens (called also Cryptogamic Plants), such as the Fungi, have, or seem to have, no leaves at all.

These spore-cases are set in clusters called *sori* (in the singular *sorus*), looking like patches of brownish or greenish brown dust, round or oblong or in lines, upon the backs or margins of the fronds ; and as no flowering-plant bears such, the full-grown Fern is easily distinguishable.

The spores — says Moore — "are minute, roundish, angular, or oblong vesicles, consisting of two outer layers, or coatings, enclosing a thickish granular fluid, and they are very numerous and arranged without order within the spore-cases. They are so small and dust-like, that when thinly scattered over a sheet of paper they are scarcely visible to the naked eye, though lying by thousands amongst the also minute emptied spore-cases. The colour, no less than the form of these spores, is variable ; they are usually pale brownish or yellowish, but they are sometimes green, and the tints of brown and yellow are much varied. These organs differ obviously from seeds, in that they consist merely of a homogeneous cellular mass. In true seeds the radicle (or young root) and the plumule (or young shoot) are present in the embryo, and are developed from determinate points ; but Fern spores, consisting merely of a small vesicle of cellular tissue — a vegetable cell,[*] grow indifferently from any part of their surface, the parent cell becoming divided into others, which are again multiplied and enlarged, until a small germinal scale, or primordial frond, is formed, and from this, in due time, the proper fronds are produced. The surface of the spores is sometimes smooth, sometimes tuberculate, or even echinate " (prickly like a hedge-hog).

From this almost invisible dust spring the multitudes of Ferns that crown the summer with their various plumes. Each atom of dust becomes a green speck, then a scale in which root and stem and leaves are yet but one confused and undeveloped mass, then a bud, then a young frond pushing its crozier-like form or its tender spikelet through the earth, then a full-grown magnificent plume like the Royal Fern—twelve feet high by the Irish Lakes, or a dainty coronal of feathers like the common Male Fern so abundant in our own English Mountain District.

The proper roots of Ferns are fibrous, proceeding from the under side of the stem when the stem is prostrate or creeping, but from all sides indifferently when it grows erect. When sufficiently numerous they form entangled masses. The fibres are mostly rigid

* Hence the name of Cryptogamic, — from crypt, or cell.

and wiry, often in youth more or less covered with fine soft hairs.

The stem is sometimes called a caudex, sometimes a rhizome. The *caudex* is the root-stock, not the root, but a true stem, either uprightly-growing or drooping, the upright stem of some foreign ferns sometimes growing to the height of fifty feet or more, like a forest tree. The *rhizome* is the creeping stem, or that part of the stem extending on or under the ground, extending very far indeed in some ferns, farthest in the Common Bracken. When not under ground, these creeping stems are generally clothed with hairs or scales, sometimes becoming quite shaggy. The rhizome varies considerably in size, from that of the Common Polypody, which is as thick as one's little finger, to that of the Film Ferns, as fine as thread.

The fronds consist of two parts, — the leafy portion, and the *stipes*, which is the part of the stalk above the caudex or rhizome. The farther continuation of the stalk, forming in the leafy part a midrib, or midvein, which becomes branched when the frond is divided (as in the Oak Fern), is called the *rachis* (*rachides* in the plural). The stipes is generally more or less furnished with brownish membranous scales, sometimes only a few at the base, sometimes extending along the rachis. When the frond is divided quite down to the rachis, or midrib, it is said to be *pinnate*, and each of the leaf-like divisions is called a *pinna* (Latin for a feather). When these *pinnæ* are again divided in the same manner the frond becomes *bi-pinnate*, or if thrice divided *tri-pinnate*. When the division is nearly but not quite down to the rib or midrib the *pinnule* (or small pinna), the pinna, or the frond, is called *pinnatifid*.

The TRUE FERNS are developed in a peculiar manner, coming up in a crozier-like form, having the rachis rolled in from the point to the base. In the more compound ferns the frond-divisions are each again rolled in after the same fashion. This is called being *circinate*. All the British species are circinate (and therefore True Ferns) except two — the Adder's Tongue and the Moonwort, in both of which the fronds are what is called *plicate*, or folded straight, like the folding of a lady's fan.

The order in which the veins, or ribs, of the fronds are disposed is called the *venation;* and deserves attention as affording one of the means of distinguishing the groups. It is from some determinate part of the veins that the spore-cases proceed. This part is called the *receptacle.* In some few native kinds the receptacle is projected beyond the margin of the frond, and the spore-cases are collected round its free extremity. More commonly, however, the veins stop within the margin, and the spore-cases grow in round or elongated clusters, situated sometimes at their ends, sometimes at their sides, and protruded through the skin of the lower surface of the fronds.

The seeds (it has already been said) are called *spores*, the seed-vessels *spore-cases*, the clusters of spore-cases *sori*. These sori, generally placed on the back or margins of the fronds, are in the great majority of British species surrounded or girt by an elastic ring or band, — sometimes vertical and burst by an irregular transverse fissure when the spores, having reached maturity, need to be dispersed,—sometimes horizontal or oblique, instead of vertical.   In the earlier stages of their growth the sori are also covered with a thin transparent membrane, called an *indusium*.   As the sori grow, the indusia get broken and thrust back, sometimes flung off.   To some species there is no perceptible indusium.   Its presence or absence therefore affords yet other means of help toward correcter classification.

What classification itself is, how far from being exact to the wonderfully arranged variety of nature, however necessary it is to assist the memory and the understanding of the botanist, may be seen from the interpenetration and confusion of characteristics even in the Table of British Ferns which follows here, and may be gathered also from a few remarks by Dr. Lindley qualifying an attempt to precisely describe only the three great primary orders of Exogens, Endogens, and Acrogens.   Having spoken of the principal differences between the three classes as to be briefly expressed thus :—

Exogens, — *wood growing concentrically — leaves with reticulated veins — flowers with their parts arranged in fours or fives—embryo, or germ, dicotyledonous (or two-leafed),—*

Endogens, — *wood confused — leaves with parallel veins—flowers with their parts in threes — germ monocotyledonous (or one-leafed),—*

Acrogens,— *wood sinuous — leaves fork-veined or altogether unbranched — no flowers — and no germ,—* he adds : —

"In applying these differences to practice, it is necessary to attend to the following rules: —

"The classes are not *absolutely* distinguished from each other by any one character, but by the *combination* of their characters. For this reason a plant may have one of the characters of a class to which it nevertheless does not belong, because its other characters are at variance with that class. Thus some species of Ranunculus have the flowers with their parts in threes; but they do not on that account belong to Endogens, because their wood is concentric, their leaves netted, and their embryo dicotyledonous. Arum maculatum has reticulated leaves; but it is not an Exogen, because its wood is confused, and its embryo monocotyledonous; its flowers are neither in fours or fives nor threes, all the parts being in a state of peculiarly diminished structure. The genus Potamogeton (a water plant, one of the Naiads) has the flowers in fours; yet it does not belong to Exogens, because its leaves have parallel veins, and its embryo is monocotyledonous."

No better words could have been written, whether to stimulate the learner to care and thoroughness in research or to rebuke the dogmatism of pedantic system-builders and teach that modesty and liberal allowance of dissent which should characterize the student of Nature and the worshiper of Truth.

FERNS — in Latin FILICES — are flowerless plants, bearing seed-vessels (spore-cases) on their fronds. "All Ferns" says Moore, — "are referrible to one of three groups: *Ophioglossaceæ — Polypodiaceæ — Marattiaceæ.* Of these the Ophioglossaceæ and Marattiaceæ are but small groups, while the Polypodiaceæ include the greater portion of all known Ferns. These three groups may each be regarded as a distinct order of plants, forming together the Filices, or Ferns." The Marattiaceæ are not found in the British Isles. We have therefore only to do with the two natural orders— POLYPODIACEÆ or *True Ferns* and OPHIOGLOSSACEÆ or *Adder's Tongue Ferns.*

POLYPODIACEÆ *are Ferns whose young fronds are rolled up in a circinate form, and whose spore-cases are girt with an elastic ring.* The presence of this ring, in some form or other, nearly or completely surrounding the spore-case, is the distinguishing peculiarity of the True Ferns. Polypodiaceæ are divided into those whose spore-cases are without valves and those which are two-valved. In the first division are two tribes or groups : — *Polypodineæ,* without valves, bursting by an irregular and transverse cleft, the elastic ring vertical and nearly complete; and *Trichomanineæ*, without valves; and bursting irregularly, but surrounded by urn-shaped involucres, the ring horizontal or oblique, complete. The second division has only one tribe, or group:—*Osmundineæ*, spore-cases two-valved, opening vertically or at the top, the ring merely rudimentary.

OPHIOGLOSSACEÆ *are Ferns whose young fronds are folded up straight, and whose spore-cases have no ring.* They are two-valved like the Osmundineæ.

POLYPODIACEÆ.

1.—Polypodineæ; 2. — Trichomanineæ; 3. — Osmundineæ.

POLYPODINEÆ—fructification dorsal, that is, the spore-cases borne on the back of the frond. Comprising the sub-groups *Polypodieæ, Gymnogrammeæ, Aspidieæ, Asplenieæ, Lomarieæ* or *Blechneæ, Pterideæ, Adianteæ, Cystopterideæ, Peranemeæ.*

POLYPODIEÆ — Sori (or clusters of spore-cases) round, and with no special indusium (or covering): comprising two genera — *Polypodium, Allosorus.*

* GYMNOGRAMMEÆ — Sori linear, no indusium : one only genus — *Gymnogramma.*

* Of GYMNOGRAMMEÆ — genus *Gymnogramma*—there is only one British species, the Small-leafed Gymnogram— *G. leptophylla*, lately found in Jersey, hardly therefore a British fern at all. Of ADIANTEÆ again its one British genus, *Adiantum*, has only one species, the Common Maiden-Hair Fern—*A. Capillus Veneris*, not found in the Lake District. Neither is the one British species of the genus *Trichomanes*, the Bristle Fern — *T. radicans.*

ASPIDIEÆ — Sori round or roundish, springing from the backs of the veins, having a special indusium : two genera — *Polystichum, Lastrea.*

ASPLENIEÆ — Sori oblong, or elongated, springing from the sides of the veins, having a special indusium : four genera — *Athyrium, Asplenium, Scolopendrium, Cetarach.*

LOMARIEÆ — Sori forming longitudinal lines between the midrib and margins of the leaflets, or divisions, of the fronds, with a special indusium : one genus — *Blechnum.*

PTERIDEÆ — Sori borne upon the frond-margins, which are changed, continuously or interruptedly, into special indusia : one genus — *Pteris.*

* ADIANTEÆ — Sori in patches on the reflexed lobes of the frond-margins, which form indusia : one genus — *Adiantum.*

CYSTOPTERIDEÆ — Sori with special oval indusia affixed behind and bent hood-like over them : one genus — *Cystopteris.*

PERANEMEÆ — Sori roundish and springing from the backs of the veins, with special involucriform or semi-involucriform indusia : one genus — *Woodsia.*

TRICHOMANINEÆ — fructification marginal, that is, having the spore-cases on the edges of the frond; sori produced around the ends of veins projecting from the frond-margins, and surrounded by urn-shaped membranes, expansions of the frond : two genera — * *Trichomanes, Hymenophyllum.*

* * See note at page 9.

OSMUNDINEÆ — fructification marginal-paniculate, that is, having the spore-cases on the edges of distinct stalks, or in irregular dense branching clusters, terminating the fronds: one genus—*Osmunda*.

OPHIOGLOSSACEÆ.

Fructification paniculate, in irregularly branching clusters, spicate, or sessile (sitting close to the stem without any sensible stalk) in two ranks on a simple spike, terminating a separate branch of the frond: two genera — *Botrychium, Ophioglossum*.

CIRCINATE

LIST OF LAKE FERNS.

(16 *Genera* — 35 *Species.*)

POLYPODIUM: — (1) P. vulgare, (2) P. Phegopteris, (3) P. Dryopteris, (4) calcareum.

ALLOSORUS: — A. crispus.

POLYSTICHUM: — (1) P. Lonchitis, (2) P. aculeatum, (3) P. angulare.

LASTREA: — (1) L. Thelypteris, (2) L. montana, (3) L. Filix-mas, (4) L. remota, (5) L. rigida, (6) L. cristata—*var.* spinulosa, (7) L. dilatata, (8) L. æmula.

ATHYRIUM: — (1) A. Filix-fœmina.

ASPLENIUM: — (1) A. Adiantum-nigrum, (2) A. marinum, (3) A. Trichomanes, (4) A. viride, (5) A. Ruta-muraria, (6) A. germanicum, (7) A. septentrionale.

SCOLOPENDRIUM: — S. vulgare.

CETERACH: — C. officinarum.

BLECHNUM: — B. spicant.

PTERIS: — P. aquilina.

CYSTOPTERIS: — C. fragilis.

WOODSIA: — W. ilvensis.

HYMENOPHYLLUM: — (1) H. Tunbridgense, (2) H unilaterale.

OSMUNDA: — O. regalis.

BOTRYCHIUM: — B. lunaria.

OPHIOGLOSSUM: — O. vulgatum.

THE POLYPODIES.

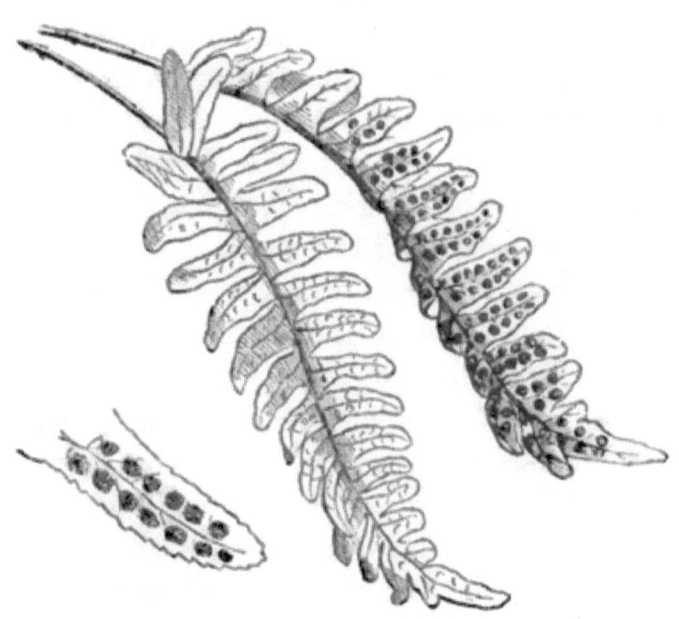

THE COMMON POLYPODY.

Polypodium vulgare. — Linnæus.

The name of *Polypodium*, meaning *many-footed*, is given to this genus, on account of the many branchings of its rhizome (or creeping stem). The common Polypody is an evergreen (the fronds, even in severe frosts, lasting till new ones are produced), growing luxuriantly on tree trunks, moist rocks and walls, and mossy banks, easily distinguished by its large round patches of orange or tawny-brown spore-cases, no other of our native ferns having its fructification at all similar in appearance.

The rhizome of the Common Polypody is often as thick as a man's little finger, covered with light brown chaff-like scales, tapering to a point, sometimes drooping so as to leave the upper surface smooth and greenish. From this upper surface spring the fronds, and from the lower side chiefly the densely-matted fibrous roots by which it clings for support. The stipes (or stalk) is naked, sometimes nearly as long as the leafy portion, the whole frond measuring from two to eighteen inches or more. The general outline is lance-shaped, very deeply pinnatifid, the lobes or segments oblong, generally round but sometimes bluntly pointed at the end, and occasionally notched along the margin. Each lobe has a slightly wavy mid-vein, or rib, branching alternately, each branch having four or five alternate branchlets, the lowest of which on the side next the point of the frond (rarely any other) produces a sorus at its club-shaped head. The fructification is usually confined to the upper part, and is generally ripe by the end of September.

The Common Polypody differs essentially from all the other British species associated with it, in having its fronds articulated with the rhizome, — that is attached in such a manner that they fall off at the approach of decay. Its texture, too, is stouter and firmer than that of other native species. The rhizome is per-

rennial. It is one of the commonest ferns, found
everywhere, on the coast line, and (in the Scottish
Highlands) at the height of 2,100 feet, very abundant
and handsome in the Lake District, abundant also
throughout Europe, and the north of Africa, found also
in Caffraria, in northern Asia from the Ural Mountains
to Japan, and widely dispersed in North America.
Its medicinal reputation is as old as Pliny, who says
that the root, dried and powdered and snuffed up the
nose, will destroy polypus. It is supposed to be the
"rheum-purging Polypody" of Shakespere; and in some
country places they still use a decoction of the fronds
as a remedy for colds and hooping-cough, employing
for the purpose the ripe fertile fronds, which, called
Golden Locks and Golden Maidenhair, are gathered
in the autumn and hung up to dry like other herbs.
The fronds contain a large quantity of carbonate of
potass. It is easily cultivated, requiring only a light
porous soil and that the rhizome should be kept on
the surface, with a constant supply of moisture, good
drainage, and moderate shade.

VARIETIES.

bifidum, — Burneside Hall, J. M. Barnes.

acuto-bifidum, — Stainton, J. M. Barnes.

auritum, — not rare.

> (Having next the rachis in the upper margin, a distinct lobule, like a
> small ear.)

crenatum, — not rare.

foliosum, — Silverdale, J. M. Barnes.

multiforme, — Whitbarrow, J. M. B.; Windermere, F. Clowes.

> (A most variable fern — no two fronds being alike.)

pulcherrimum, — Whitbarrow, **J. Addison.**
  (Like semilacerum, but more beautiful.)
serratum, — in several places.
submarginatum, — Levens, **J. M. Barnes.**
marginatum, — Windermere, **F. Clowes;** Arnside, **J.** Crossfield.
truncatum, — Crosthwaite, **F.** Clowes; Levens, **J. M. Barnes.**
variegatum, — Witherslack, **J. M. Barnes.**
multifidum, — Staveley, Martindale, Whitbarrow, **F. Clowes.**
semilacerum, — Whitbarrow, **J. M. Barnes, F. Clowes;** Ulver-
  stone, **J.** Crossfield.
semilacerum robustum, — **Whitbarrow, J. M. Barnes.**
hastifolium, — Witherslack, **M. Read.**
sinuato-auritum, — **Whitbarrow, J. M. Barnes.**
Whitakeri, — near Kendal, **M. Whitaker.**
  (**A** very uniformly crenate form.)

VAR. BIFIDUM

MOUNTAIN POLYPODY,

OR BEECH FERN.

Polypodium Phegopteris.— LINNÆUS.

The Beech Fern is one of the tenderer
ferns : produced from the perennial rhizome about May, and dying off in the
autumn or at the first approach of frost.
It grows abundantly on the slate in moist mountainous places and in the shade of damp woods, from
the coast level to (in the Western Highlands of
Scotland) an elevation of over 3000 feet. The stem is
slender, creeps very extensively, and is slightly scaly,

c

producing black fibrous roots. From it spring delicate hairy pale-green fronds, to the height, when full-
grown, of from six to twelve inches. The stipes is
fleshy and very brittle, frequently longer, sometimes
much longer, than the leafy part of the frond, having
near its base a few small and almost colourless scales.
The fronds are triangular, extending to a long narrow
point; in the lower part pinnate,— but with this division seldom carried beyond the two lowest pairs of
branches, those of the upper parts of the fronds being
pinnatifid (connected at the base). The pinnæ (or
leaflets) have an acutely lance-shaped outline and
are deeply pinnatifid, usually standing in pairs, opposite to each other, the lowest pair drooping toward the
root and set on at a short distance from the rest. The
united bases of the pairs of the other pinnæ — when
they happen to stand exactly opposite to each other —
exhibit more or less distinctly a cruciform figure, by
which, regard being had also to the general triangular
outline and sub-pinnate division, this species may be
known from the other Polypodies. The venation consists of a slender flexuous midvein (or rib), from which
proceed alternate or sometimes opposite veins extending to the margins of the lobes or lobules,— either
simple or once forked at about half of their length.
The veins when simple, or the anterior venules when
divided, bear a sorus at a short distance from the edge
of the lobule. This almost marginal fructification
extends nearly over the whole frond, the sori being
small and circular, arranged in series, and often becoming confluent in lines. When the fructification is but

partially developed only one or **two of the** lowermost veins **are** fertile, in **which case** the marginal series of **sori is not** very manifest. The spore-cases **are** small, numerous, and of a pale brown hue; the spores ovate and smooth. The **fronds in this** species become lateral and distant from **each** other on the **underground stem,** in consequence **of its** rapid elongation; and they are adherent, **that is** to say, the stipes is not **furnished with a natural** point **of spontaneous** separation.

For cultivation in **pots or** cases it requires good **current** moisture, and **grows** well **on the** damp and shady sides of garden rock-work.

HABITATS. — **Scawfell,** Wastdale, **Borrowdale, En-** nerdale, **Keswick, Tindal Fell,** Newbiggin Woods, Stockghyll **Force, Ambleside, Grasmere,** Casterton Fell, Wallington, **Coniston, &c.**

VARIETIES.

interruptum, — Witherslack, J. M. Barnes.
multifidum, — Burton, J. Jones.

SMOOTH THREE-BRANCHED POLYPODY
OR OAK FERN.

Polypodium Dryopteris.*—— LINNÆUS.

This, the smallest of the Polypodies, is also one of the most delicate of all British Ferns: very easily recognizable by its smooth fronds, of a bright lively green, divided into three branches,— the last characteristic even more obvious in the young fronds, which are rolled up in little balls at the ends of their three slender stalklets. Its height is generally not more than six inches, often less, but it sometimes grows to twelve or fourteen. It is fragile, produced about April, and in succession throughout the summer, soon withered by heat or drought, and at once destroyed by frost. The fronds rise from a slender creeping stem, which often forms densely matted roots. The stipes is usually much longer than the leafy part, thin, brittle, and dark coloured. The general outline is five-sided, owing to the division of the fronds into three triangular branches. One of the peculiarities of the Oak Fern is the deflection of the rachis (or midrib) at the point where the branches take their rise; and another (of less botan-

* Also Polystichum Dryopteris, Lastrea Dryopteris, &c.

ical importance, but very helpful in distinguishing it from its near ally, the Lime stone-Polypody) is its perfect smoothness, a constant distinction, most easily seen on the stipes and rachis, but equally occurring over the whole plant. The fronds are divided so that each branch is pinnate at the base, pinnatifid toward the point; the pinnæ are also pinnate at their bases, then pinnatifid, and at their points acute and nearly entire; the pinnules and ultimate lobes are oblong and obtuse, with a rather wavy midvein, from which the venules branch alternately, extending to the margin,— in those of moderate size simple with a sorus at each extremity, in the larger branched with the sorus on the lower branch. The fructification varies much, according to habitat and season, being sometimes very much crowded and sometimes very sparse. The spore-cases are small, roundish obovate (inversely egg-shaped), and attached by a slender stalk ; — the spores ovate (egg-shaped), oblong, or roundish, with a granular surface.

Like the Beech Fern, the Dryopteris is found from the coast-level to a great elevation. It is seen

sometimes **on** the lime-stone with the Lime-stone
Polypody, **but** is rather to be **classed as a** slate fern,
liking a rocky district, with running water, and the
not stagnant moisture of woods, needing both shade
and shelter. It **is easily** cultiveable in **a mixture of**
fibre, peat, leaf-mould, and sand, either in the house **or**
on artificial rock-work.

HABITATS.—— Lodore, Borrowdale, Honister Crags,
Scale Force, Gillsland, Wastdale, Dalegarth, Stock-
ghyll, Glenridding, Hutton Roof, Casterton, Coniston,
Furness Fells, &c.

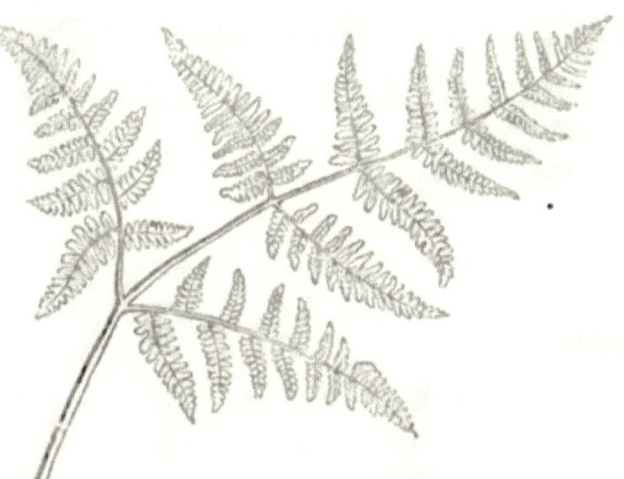

THE LIME-STONE POLYPODY.

Polypodium calcareum.* — SMITH.

The Lime-stone Polypody differs from
the Oak Fern in its stouter, erecter, and
more rigid habit, and in the mealy-looking
glandulosity of its surface; yet still more
in that the division of its fronds is pinnate
rather than into three distinct branches.
The fronds are also of a duller and deeper
green, and without the marked deflection
of the rachis. And instead of the branches being
rolled up into three balls, the young pinnæ curl in on
their rachides and the entire frond upon its rachis, so
that the frond is of the ordinary bipinnate structure.

* **Also P.** Robertianum, P. Dryopteris, Lastrea calcarea, Lastrea
Robertianum, Phegopteris calcarea, &c.

Its fronds, including the stipes, vary from six to eight-
teen inches in height; their form is triangular with
a tendency to the pentagonal appearance of the Oak
Fern, because of the larger size of the two lower pinnæ.
These lower pinnæ are pinnate, with pinnatifid pinnules,
the upper pinnæ also pinnate, with the lower pinnules
again pinnate and the upper pinnatifid. The fructification is scattered over the whole dorsal surface of the
frond; the sori are small and round, consisting of numerous crowded spore-cases, entirely without indusia,
arranged in linear sub-marginal series along each side
of the lobules, or in series between the midrib and
margin when the lobules are but slightly developed,—
often more or less confluent. Spore-cases pale brown,
roundish obovate, small, and numerous. Spores ovate
or oblong, somewhat granular.

The Lime-stone Polypody, as its name infers, is found
usually on exposed rocky limestone tracks: growing there very abundantly. In cultivation, however, a
lime-stone soil is not essential to its well-being. Like
the generality of ferns, it requires good drainage; but
bears the sun more than most others. It grows almost
throughout Europe, in Canada, and the United States
of America; and has been gathered by Dr. Hooker and
Dr. Thomson on the Himalaya Mountains at the elevation of 6,000 feet.

HABITATS. — Scale Force (*J. Robson*), Whitbarrow,
Newbiggin Woods, Gilt Quarries, Baron Heath, Arn-
side Knott, Hutton Roof Crags, Farleton Knott,
Caskill Kirk, &c. Only one variety:

variabile, — Whitbarrow, J. M. Barnes and F. Clowes.

THE ROCK BRAKES.

THE MOUNTAIN PARSLEY FERN.

Allosorus crispus.— BERNHARDI.
Osmunda or Pteris crispa.— LINNÆUS.

Of the Rock Brakes there is but one British species —
the Mountain Parsley Fern, known at once by its likeness to tufts of parsley, and distinctly differing from
other of our native ferns in the marked division of its
sterile and fertile fronds,— the first of which have these
segments broad, flat, and leaf-like, while the second
have them involute, or rolled in at the margin, covering the sori instead of an indusium. The fronds of the
Mountain Parsley Fern are annual, coming up in May
or June, and dying down in the autumn, from four to
twelve inches high (including the stipes), of a lively

green, triangular or ovately triangular in outline. The barren fronds are generally as long as the stipes, bi or tri-pinnate, and smooth. The segments or leaflets into which they are cut are more or less wedge-shaped and notched or cleft at their ends. The fertile fronds have the leaflets of an oval or oblong or linear form. The venation of the barren fronds consists of a slender vein extending along each pinnule, casting off another into each of its lobes, this again alternately branching, one branch being directed towards every marginal point. In the fertile fronds a midvein enters each ultimate division and passes sinuously to its point, throwing out nearly to the margin alternate veins, usually simple but sometimes forked, bearing a sorus near their ends. The fructification usually occupies the whole system. The sori small, roundish, at first distinct though contiguous, ultimately becoming laterally confluent and forming a continuous line. Spore-cases small, elliptic obovate, stalked. Spores smooth, roundish, oblong or bluntly triangular.

The Mountain Parsley Fern is peculiarly a mountain plant, delighting in the shades and corners of boulders, and to be among loose slate-stones, and at the feet of the unmortared walls that wind about the fells and mountains. It is well-fitted therefore for garden rockeries; but is apt to die off in winter if allowed to be too damp.

HABITATS.— Skiddaw, Keswick, Whinlatter, (*W. Christy*), Borrowdale, Ennerdale, Scawfell, Helvellyn, Blencathra, Kirkstone Pass, Ambleside, Coniston, Grasmere, &c., &c.

THE SHIELD (AND BUCKLER) FERNS.

THE ALPINE SHIELD FERN,
OR HOLLY FERN.

Polystichum Lonchitis.*—ROTH.

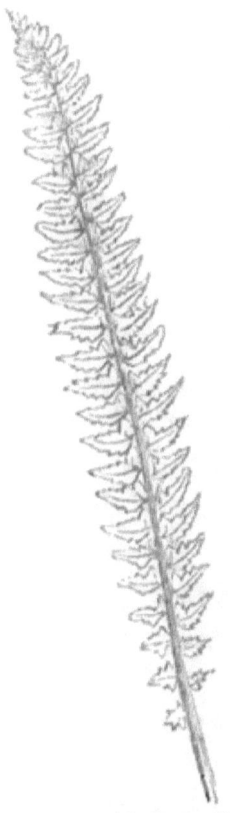

POLYSTICHUM is an extensive genus of the group of Aspidieæ; and consists for the most part of harsh spiny evergreen ferns, scattered from the torrid to the frigid zone, —but represented in this country by only three species : — *Polystichum Lonchitis, Polystichum aculeatum,* and *Polystichum angulare.* These species, however, are very variable, and so closely connected by intermediate forms that it is difficult sometimes to distinguish their exact limits.

The Alpine Shield Fern is one of the rarer ferns, taking its other name of Holly Fern from its hard and prickly appearance. It is an evergreen, with a

* *The Polypodium Lonchitis* of Linnæus.

scaly tufted stem from the crown of which the young fronds are produced in spring, to remain fresh and vigourous until the spring following, arriving at maturity in autumn and flourishing through the winter. It is a true rock fern, growing to a height of from six to nine inches, but higher and more luxuriantly in damp and slightly elevated situations. The fronds are of a deep green, pale beneath, of a rigid leathery texture, erect or drooping according to the conditions of their growth; once pinnate, and in their general outline narrowly lance-shaped, or lanceolate. The pinnæ are short, crowded, and shaped something between a sickle and a crescent, the upper side of the base having an ear-like projection, called an auricle, while the lower side is as if cut away. The margin is set with spinous teeth. The venation is very indistinct, the veins twice-branched, the branches extending to the margins without joining with others. The sori form a line on each side the mid-rib, parallel with it and half way between it and the margin, generally becoming confluent in age, and covered by a circular membranous scale attached by a short central stalk. Spore-cases deep brown. The name Polystichum is from two Greek words signifying *many* and *order*, given on account of the number and regularity of the lines of sori. It is difficult of cultivation, needing pure mountain air, and therefore seldom thrives under artificial treatment.

HABITATS. — Helvellyn (*Isaac Huddart* and *F. Clowes*), Fairfield (*James Huddart*), Deepdale (*M. Hankey*), Ullswater (*Rev. W. H. Hawker*), Farleton Knott (*J. Jones*).

THE COMMON PRICKLY SHIELD FERN.

*Polystichum aculeatum.**—ROTH.

The Common **Prickly Shield Fern** is one **of the
larger** and hardier **ferns**, preferring, however, **a loamy
soil** and the partial shade **of woods or** hedge-banks,
where it grows to the height **of from a** foot to **two feet
or more**, with a short stipes **densely** enveloped in rust-
coloured membranous pointed scales.   The fronds, from
four to seven **inches** across, **are, like the Alpine** Shield
Fern, rigid and leathery in **texture**, of a shining dark
green above, paler beneath, **erect and spreading**, or
occasionally **drooping, growing up in a circle in April**
or May, **from a stout** tufted **stem, or crown.   The**
general typical **form is** broadly **lanceolate ; but the**
variety LOBATUM is very narrowly lanceolate ; bi-pin-
nate, with **alternate** pinnæ, these **pinnæ** being again
more **or less** divided into a series of pinnules, either
decurrent — that is insensibly **merging in** the substance of the rachis which **supports them**, or tapering
to a **wedge-shaped base and attached to the rachis**
by **the point.   The pinnules are of** a long **crescent**
shape, with the upper **base** extended **into a** small
auricle, **or** enlarged lobe, **and** the **lower** base sloped
away,— the apex going off to an **acute point**, and the
margin notched **with spiny teeth.   Venation, fructification, and indusium, similar to** P. *Lonchitis.*

* *Polypodium aculeatum,*— Linnæus. Also *Aspidium aculeatum.*

The variety LOBATUM, considered a distinct species by some botanists, differs chiefly in the narrower outline of the frond (already noticed), and in the pinnules being much more decidedly decurrent, or running together, at the base. Every possible variation in the relative portions of the pinnules is to be met with, from the typical bi-pinnate form of *Polystichum aculeatum* to a simply pinnate form of the species, which from its resemblance to *P. Lonchitis* has been called *lonchitidioides*. This latter, owing its origin only to some special circumstances of its growth, cannot be considered as a permanent variety; but the intermediate variation — *lobatum*, which is the most common of these abnormal ferns, is at least sufficiently different to be considered a variety.

The Common Prickly Shield Fern is one of the most easily cultivated of all our larger and hardier species, while the *lobatum*, being smaller, is well adapted for pots. With good drainage and moderate shade, they thrive admirably.

HABITATS. — Irton Woods (*Robson*) — Ara Force (*H. Fordham*)—Ambleside and Rydal.

VARIETIES.

grandilobum, — Mardale, J. M. Barnes.
multifidum, — Levens, J. Wood; Whiteside Fell, J. M. Barnes;
    Silverdale, J. Crossfield.

THE SOFT PRICKLY SHIELD FERN.[*]

Polystichum angulare. — PRESL.

This specimen is not easily distinguished from
P. aculeatum, though certainly distinct. The two
may, however, be generally known from each other by
the following differences : — 1 — *P. angulare* is less
stout, less erect, and altogether less rigid in texture,
normally lax and more herbaceous, while equally large
or larger. 2 — *P. aculeatum* has its pinnules either
confluent or decurrent (in which case there is no diffi-
culty in distinguishing it), or when the pinnules are
distinct, as in the most perfect plants, they are wedge-
shaped at the base, the anterior side being truncate,
and the posterior obliquely incised in straight lines, the
two lines describing an acute angle, by the apex of
which they are attached to the rachis ; while in *P.
angulare* the truncated anterior base is more curved in
outline and the two lines of the base describe a right
angle or an obtuse angle, at the apex of which is a distinct slender stalk, by which they are attached. 3 —
P. aculeatum has its sori attached at a point along
the middle part of the venule, the apex of which is
carried out to the margin of the pinnule, the sori thus
being placed nearer to the point of forking or branching
than to the apex of the venule ; while in *P. angu-*

---

[*] Or Angular-lobed. Polystichum aculeatum, Aspidium angulare, A. aculeatum, Polypodium angulare, &c.

lare the fertile venule stops about midway across the
pinnule and the sorus is generally placed at or almost
close to the apex.   The basal pinnules and the portion
rather below the middle of the frond should be taken
for examination.   The upper parts of the fronds alone,
in these Polystichums, are useless for the purpose of
identification.

COMMON AND PRICKLY SHIELD FERNS.

The Soft Prickly Shield Fern is one of our most
graceful ferns, strong-growing and tufted-stemmed,
sometimes forming great masses, the fronds lanceolate
and rising to the height of from two to five feet, lasting
through ordinary winters and in sheltered places re-
taining their verdure until the new fronds are produced,
the old fronds only gradually dying off as the new
ones become developed, in or about May.   The stipes,
varying from a third to a fourth of the length of the

whole frond, is very shaggy, with reddish chaff-like
scales continuing though decreasing in size throughout
the upper portions of the frond. The fronds are bipinnate, with numerous tapering distinct pinnæ, having
their pinnules flat, and somewhat crescent-shaped, from
the prominent auricle at the anterior base, often bluntish at the apex but sometimes acute, always with
spinulose marginal serratures, and sometimes in a few
of the lower pinnules with deep lobes so that the pinnules become pinnatifid. The pinnules taper to an
obtuse or right-angled base, and are attached, as before
said, by a slender stalk, which does not form a line with
either margin. The pinnules have branched free veins;
and the sori are generally ranged in a row on each side
of the midrib and covered by a peltate (fixed to the
stalk by the centre) scale or indusium.

Not only one of the most beautiful, this is also
one of the most remarkably varied of our Ferns.
Evergreen, and able to readily accommodate itself to the
changes of artificial culture, it is specially fitted for the
out-door or in-door fernery; growing readily in pots
(with sufficient room) in the garden or shrubbery, in
free sandy loam, or on shady rockwork. It is easily
increased by division whenever lateral crowns are produced. It is rare in the North of England, or of
Europe, though found in Sweden and Norway; but is
more plentiful in the South of England, and very
abundant in Central and Southern Europe; in Asia
also, from Georgia to India and Ceylon; in Abyssinia
and on the African coast of the Mediterranean; and in
North and South America, in New England and in

Mexico and Chili. It is found not infrequently in the Lake District, generally in warm sheltered ghylls.

VARIETIES.

biserratum, —Whitbarrow, **J. M. Barnes;** Beetham, **J. Crossfield.**
acutum dissectum, — Whitbarrow, F. Clowes.
proliferum, — Humphrey Head, **A.** Mason.
tripinnatum, — Arnside, **J. Crossfield.**

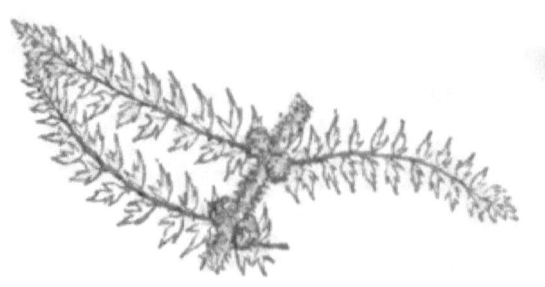

VAR. PROLIFERUM

THE FEMALE BUCKLER FERN,

OR MARSH FERN.

Lastrea Thelypteris.* — PRESL.

The *Lastreas* (not Lastræas as often written) take
their name from M. deLastre. They are to be known
from *Polystichum* (both belonging to the Aspidieæ)
by the outline of the indusium, which is kidney-shaped,
or roundish with a notch in the side, the attachment
to the frond being at the notched part. There are no
less than eight species of this genus found in Great
Britain (and in the Lake District), and the group in-
cludes some of the largest, the commonest, and the
most elegant of our ferns.

The name of Marsh Fern (sometimes Marsh Buckler
Fern) comes of course from the place of growth, peaty
bogs or marshy land.

The rhizome creeps extensively, sparingly branched,
producing fronds at intervals, slender, smooth and
blackish brown, having a few pale brown scales at the
growing point, and numerous fibres. The stipes is as
long as or longer than the leafy portion in the fertile
frond, rather shorter and slighter in the barren, the
whole frond growing to the height of from six inches
to more than three feet, the fertile fronds the taller.

---

*The Polypodium **Thelypteris** of Linnæus, also Aspidium The-
lypteris, Polystichum Thelypteris, Athyrium Thelypteris, &c.

The fronds, produced about May and perishing in Autumn, are delicate in texture, pale green, lanceolate, and pinnate; the pinnæ mostly opposite, a short distance apart, and pinnatifidly divided into numerous crowded entire rounded lobes, the lobes of the fertile frond appearing narrower and more pointed on account of the bending under of their margins. The venation of the lobes consists of a distinct somewhat wavy midvein, from which alternate venules branch or fork out, each branch bearing a sorus half way between the midvein and the margin. The sori often become confluent and are partly concealed by the bending back of the margin. The indusium is small, thin, shapeless, and soon thrown off.

The Marsh Fern is to be known from the other Buckler Ferns by its long, comparatively slender rhizome, which is unlike that of any other native species. It ought not therefore to be confounded, as it sometimes is, with *L. Oreopteris* which has a short thick tufted caudex, merely decumbent in habit. It differs farther in having its fronds of their full width almost to the very base, and supported by a long bare stipes, while *L. Oreopteris* has diminishing pinnæ carried down almost to the base of the stipes; and in its fronds being almost free from glands, while those of *L. Oreopteris* are very conspicuously glandular and very fragrant. It is still less like other species. It is readily cultivable, wanting only a moist peaty situation, a damp garden border or a boggy pool, where its rambling stems may have room to spread. If set in pots they should be large and shallow.

THE

MOUNTAIN BUCKLER FERN,

SWEET MOUNTAIN FERN, OR HEATH FERN.

*Lastrea Oreopteris.** — PRESL.

The Sweet Mountain Fern is known at once by its balsamic scent, the fragrance of which is given out strongly from numerous minute resinous glands on the lower surface when the frond is drawn through the hand. The fronds are noticeable also for their coronal appear-

_ *Lastrea montana (Moore), Polystichum Oreopteris (deCandolle), Aspidium Oreopteris, A. odoriferum, Polypodium montanum, P. fragrans, P. Thelypteris, Phegopteris Oreopteris, &c.

D 3

ance, set on the stem like the feathers of a shuttlecock
and growing in graceful tufts to two or three feet high.
They are annual, springing up in May and dying off
in Autumn; bright green or yellowish, erect, lance-
shaped in general outline, and pinnate.   The stipes is
unusually short, the leafy part being continued nearly
to the ground, and the lower pinnæ becoming so short
that the frond tapers downward as much as toward
the point.   The pinnæ are generally opposite, narrow,
tapering and pinnatifid, and bear the sori almost close
to the margins, in most instances very abundantly.
The fronds differ, as was said, from *L. Thelypteris* in
in the shortness of the lower pinnæ, and again in the
margins being flat and not turned back upon the sori.
Each lobe has a distinct and slightly wavy midvein,
alternately branched, the branches simple or divided,
with the sori near their extremities.   The indusia are
small and soon perish or fall away, sometimes seem
to be altogether wanting; but the plant is too
closely allied to other species of the genus to allow
of its separation on this account.   It grows in damp
woody places, especially luxuriating by the sides of
shady becks and waterfalls; but is much more pro-
fusely met with on heathy mountain sides. It is the
common fern of many parts of the Scottish Highlands,
growing sometimes at an elevation of 3,000 feet;
abounds in the English Lake Country, and in Wales;
and is more or less plentiful in waste districts throughout England. Though so common it seems not easy of
cultivation, needing perhaps the pure air and ready
drainage of its native mountains. In smoky London

the plants will not long retain their vigour.   It grows
abundantly throughout the district.

VARIETIES.

caudata, — Windermere, F. Clowes.
cristata, — Windermere, J. Huddart
interrupta, — Windermere, J. Huddart; Levens, Crossfield.
breviloba, — Mardale, J. M. Barnes.
curvata, — Farleton Knott, J. M. Barnes, J. Jones.
subcristata, — Mardale, J. M. Barnes.
truncata, — Potter Fell, J. M. Barnes; Farleton Knott, J. Jones

VAR. CRISTATA

THE

COMMON BUCKLER FERN,

OR MALE FERN.

Lastrea Filix-mas.* — PRESL.

The Male Fern is so called from its robust appearance in contrast with the more delicate though similar Lady Fern (Filix-fœmina). It is an annual except in sheltered spots, where the old fronds will continue green until the new come out. One of the commonest of our Ferns, it is yet one of the most beautiful, especially on account of its coronal growth, like a circle of erect but gracefully waving feathers springing from one stem, the fronds

---

* **Polypodium** Filix-mas (*Linnæus*), Aspidium Filix-mas, &c.

smooth, of a lively green, somewhat paler underneath, averaging a height of from two to three feet, but varying from twelve inches to even six feet, according to age, variety, or locality. The stipes is short, stout, and densely scaly. The fronds are broadly lance-shaped; bipinnate, though less decidedly so than some other species, only those pinnules which are nearest the main rachis being quite separate; the pinnæ are narrow and tapering, with a few of the lowest pinnules distinct, the rest united at the base, — these pinnules of an obtusely oblong form and serrated, or notched, on the margin. The sori are usually confined to the lower half of the pinnules, but they are very crowded and abundant.

This is one of the best of our Ferns to be studied in order to understand the fructification, for here the indusium, in almost fully developed fronds, is remarkably prominent, closed over the spore-cases and seen to consist of a lead-coloured tumid kidney-shaped scale, which in due time is raised on one side for the dispersion of the spores. This may be seen by watching the fronds just as they are reaching their full growth. The veins also of this species are very manifest, each pinnule having a flexuous midvein, with alternate venules, simple, forked, or sometimes three-branched in different parts of the pinnule, the three-branched occurring at the base and the unbranched at the apex. The sori are borne on the branch toward the apex of the pinnule and form a line of dots on each side of the mid-vein and at a little distance from it.

The variety L. INCISA is a magnificent Fern, growing

much larger than the commoner typical form of the plant, with the same general features, only larger in every part, its pinnules more elongated and tapering more deeply cut along their margins, the branchings of its venules more numerous, and its sori covering a larger surface, reaching almost to the apex of the pinnules. Another variety, L. PALEACEA is remarkable for the abundant and usually golden scales clothing its stipes and rachis. This variety is very distinct and permanent.

The typical Male Fern, also its Incised and Golden-scaled varieties, may be found, though not at any very great elevation, over the whole of British ground and throughout Europe from Scandinavia to the Isles of Greece. In Asia it extends from the Caucasus to Lake Baikal, and from the Ural Mountains to the Himalayas and to Assam. It is found also in Northern Africa and in Madeira. And in the Americas, in Newfoundland and in Mexico, from California to Peru and Brazil. But, it is said, not in the United States. Its culture is not at all difficult; it will grow in any shady places, in almost any kind of soil, the best a sandy loam, moist but not wet.

The Male Fern has long had and still retains a medicinal reputation as a specific against tape-worm. Galen used it; Pliny also, who also called it *Filix-mas*. Its astringent stems have been employed in tanning leather, and its ashes in bleaching linen, and making glass and soap. Bishop Gunner speaks of the young curled fronds being boiled and eaten like asparagus, and says that the poor Norwegians cut off the succulent laminæ at the crown of the root (the bases of the future stalks) and, adding a third portion of malt, brew from them a kind of beer. In times of great scarcity they mix them with their bread. Cut green and dried in the air, this Fern, like the Bracken, is used in Westmorland and Cumberland as litter for the cattle; and if steeped in hot water would, it is said by the bishop, be a not despised but readily-eaten and fattening food — for the cattle as well as the Norwegians. The young crosier-like stems were of old, called St. John's hand or "lucky hands," considered to be preservative from witchcraft.

VARIETIES.

paleacea, — Common.

abbreviata, — Common.

———— *erosa,* — Keswick, Miss Wright; Troutbeck, F. Clowes.

———— *cristata,* — Borrowdale, J. D. Harrison.

———— *interrupta,* — Windermere, F. Clowes.

Pinderi, — Elterwater, Miss Beevor.

Clowesii, — Troutbeck, F. Clowes.

Barnesii, — Levens, J. M. Barnes.

digitata, — Burton, J. Jones.

grandiceps, — Burton, Wearing.

excurrens, — Ravenscar, J. A. Martindale.

attenuato-multifida, — Mardale, J. M. Barnes.

crispata, — Levens, J. M. Barnes.

multiformis, — Long Sleddale, J. M. Barnes.

pulchra, — Mardale, J. M. Barnes.

stricta, — Mardale, J. M. Barnes.

serrata, — Burton, J. Jones.

producta, — Arnside, J. Crossfield.

THE REMOTE BUCKLER FERN.

Lastrea remota. — MOORE.

This Fern — the *Aspidium rigidum or remotum* of
Braun, makes its first appearance in England in the
Addenda to Mr. Moore's "Nature Printed British
Ferns" published in 1863, having been discovered in
the neighbourhood of Windermere, by Mr. F. Clowes
and Mr. Huddart in 1856. It had previously been
only known as a native of Southern Germany. Braun
inclines to regard it as a mere divided form of the

Common Male Fern, but Moore claims for it the rank
of a species. In general character and aspect it much
resembles those vigorous examples of *L. cristata* var.
spinulosa which are sometimes met with, having the
same narrow elongate erect fronds; but in its struc-
tural characteristics it differs materially from that
plant and agrees much more closely with *L. Filix-mas*.
From *L. Filix-mas*, the incised pyramidal-pinnuled
forms of which most nearly resemble it, it must however
be separated on account of the farther divisions of its
fronds, which are tripinnate. In *Filix-mas* also the
serratures of the lobes are often acute, while in *remota*
they often terminate abruptly in a hard short point,
being what is called mucronate, still oftener mucronulate,
thus forming the intermediate step between *Filix-mas*
and *L. dilatata*, which is mucronate and spinulose.
Mr. Clowes, who has carefully cultivated and observed
it, considers it to be a hybrid between *L. Filix-mas*
and *L. cristata* and *spinulosa*.

The caudex of *L. remota* is stout and ascending,
with a thick scaly crown; its stipes, a foot long, is
stout, and clothed with numerous scales of various
size, some ovate-acuminate, three-quarters of an inch
long, others smaller, lanceolate or linear terminating in
a lengthened hair-like point, the margin slightly wavy
or toothed, — along with these larger scales numerous
others occurring, minute, ovate caudate, and peltately
attached; the rachis, both primary and secondary, is
furnished with scales which become smaller upwards.

The fronds, from three to four feet high, are erect,
narrow oblong lanceolate, sub-tripinnate, and smooth;

the lower pinnæ three to four inches long, ovate-
acuminate, the central six inches long, linear oblong
acuminate, all ascending, opposite or sub-opposite and
distant below; pinnules (basal ones of second pair of
pinnæ) an inch and a quarter long, shortly petiolate,
pyramidal or pyramidal ovate, acute, pinnatifid, almost
pinnate; lobes oblong, about three-quarters of an inch
long, obtuse, the lowest sub-lobate at the base, otherwise serrated, the serratures most numerous and prominent at the apex, acute, mucronulate. The pinnules
become gradually less pyramidal or ovate, and more
oblong, at length linear oblong, as they recede from
the main rachis; below, except the lowest, they are
also sessile, narrowly attached, but gradually more and
more adnate upwards. The pinnules of the upper
pinnæ resemble the smaller pinnules of the lower.

The venation consists in the larger lobes of a flexuous
primary midvein from which alternate venules proceed
toward the serratures, sometimes branched, the sori
being situated midway on the simple venules and close
above the fork upon the branched. In the smaller
pinnules the vein bears a sorus midway on its lowest
anterior venule, so that a row of sori is formed on each
side of, and at a little distance from, the principal vein.
The basal lobes often bear two or three other sori, and
are traversed by a series of alternate simple venules.

The fructification occupies the whole back of the
frond, from base to apex. The sori are prominent and
distinct, in two series, near the vein of the smaller
pinnules and on the lobes of the larger. Indusium
persistent, reniform, indistinctly erose-dentate (irregu-

larly toothed) on the margin, not glandular. Spore-cases roundish obovate. Spores oblong, granulated, the majority abortive.

No known varieties.

THE RIGID BUCKLER FERN.

Lastrea rigida. — PRESL.

This Fern is **of moderate size, growing** from a foot to two **feet in** height, **erect, and spreading,** the fronds **annual,** springing from the **crown of a** comparatively **thick scaly** tufted stem, or caudex. **It is one of** the **most elegantly** divided of the *Lastreas,* **the pinnules being all doubly** and very evenly serrated, or **toothed.** The **fronds are** narrowly triangular, rarely somewhat lanceolate, bipinnate, with narrow tapering pinnæ; comparatively small, and generally broadest **at the base, always covered with minute glands, giving off a** pleasant **balsamic** fragrance **when** bruised, **to be** smelt **also in** the sunshine from **the** untouched **plants.** The outline of **the** pinnules, bluntly oblong **with** shallow **lobes** (differing in this from the other native species of **the genus), is** most nearly approached **by** some forms **of** *Filix-mas incisa,* and the serratures also, as in that, are not at all **spinulose or** bearded, but short and merely **acute (it is, however,** distinguishable from that by **its size, its outline, its** glandular surface, and its glandular-fringed indusium). **It can** hardly be mistaken **for any other** of the *Lastreas,* nearly all the rest of them having spinulose serratures.

The stipes is densely scaly. The **venation** is similar to that of *Filix-mas*, the pinnules having a flexuous midvein, with **alternate venules** again **pinnately** branched. The **sori are** borne on the lowest anterior branch of each **venule**, that is, on the lowest veinlet on the side next the **apex** of the pinnule, and are covered by a kidney-shaped indusium which does not fall away.

The Rigid Buckler Fern is almost entirely confined to a few limestone craggy tracts within a small area of the contiguous parts of Westmorland, Lancashire, and Yorkshire. The Rev. G. Pinder writes :—" I met with *Lastrea rigida* in great profusion along the whole of the great scar limestone district, at intervals between Arnside Knott (where it is comparatively scarce) and Ingleborough, being most abundant on Hutton Roof Crags and Farleton Knott, where it grows in the deep fissures of the natural platform, and occasionally high in the clefts of the rocks ; it is generally much shattered by the winds, or cropped by the sheep, which seem fond of it. With regard to the shape of the frond, I may mention that among some hundreds of specimens I found but one or two which had the fronds oblong-lanceolate, all being more or less triangular, and not having the lower pair of pinnæ shorter than those in the upper and middle parts of the fronds. The fronds of young plants are remarkably triangular. The two forms of fronds no doubt depend upon the situation, whether sheltered or otherwise, and on other causes ; still I imagine the triangular to be the true form of the plant." Its elevation above the sea appears to range between 200 and 1,500 feet. In cultivating it, it should be borne in mind that it is more impatient than other kinds of stagnant moisture, and that it is better for the caudex to be a little above the soil, to provide a better fall for the decumbent fronds.

HABITATS. — Arnside Knott, Hutton Roof Crags, Farleton Knott, Silverdale, Whitbarrow.

VARIETIES.

polyclados, — Farleton Knott, J. M. Barnes.
interrupta, — Arnside, J. Crossfield.

L. RIGIDA.

THE CRESTED BUCKLER FERN.

Lastrea cristata.* — PRESL.

Lastrea cristata, L. uliginosa, **and** *L. spinulosa,* constitute a group distinguishable by habit and other characters from the allied *dilatata* group, with which the more highly-developed form *spinulosa* is sometimes associated. "In our 'Handbook of British Ferns (2nd ed.)," says Moore, "this group was treated as consisting of three forms of one not very variable species; and notwithstanding that many Fern authorities do not appear to adopt this view, we have no doubt whatever that the plants possess a close natural affinity, and have characters which separate them from the forms of *Lastrea dilatata,* however similar to the latter, in some cases, may be the degree and mode of division in the fronds — points on which botanists are at times too prone to rely. The close affinity of the three forms now alluded to is evidenced by marks far more important than those to be derived from such characters as the outline or cutting of the fronds, namely, by the creeping caudex, by the erect narrow fronds, by the sparse and pallid broad appressed scales of the stipes, and by the entire indusia, in all which

---

* Polypodium cristatum (*Linnæus*).

respects they perfectly agree. On the other hand, it is in these points that they differ from the *dilatata* group. In the folio edition of this work, we were led, in deference to the more commonly received opinion, to treat of *spinulosa* separately; but after some years' further observation, we revert to our former view, and place it here under *cristata*."*

To this it may be well to add the special distinctions which characterize the whole group of what were once called Crested Shield Ferns — *L. cristata, uliginosa, spinulosa, dilatata,* and *æmula,* although *L. cristata* is only known in the Lake Country by its variety *spinulosa,* — the true Crested Buckler Fern and var. *uliginosa* occurring only in more southern counties.

Lastrea cristata grows with very erect, narrow, oblong fronds, whose deltoid pinnæ are not quite divided down to the midrib, the lobes being attached by the whole width of their base, and oblong, with a rounded apex. The stipes is sparingly furnished with broad, obtuse, membranous, whole-coloured scales; and the caudex is creeping.

Lastrea uliginosa has two or three sorts of fronds. One set, the earlier barren ones, resembles those of *cristata,* the fertile being bipinnate at the bases of the pinnæ; the fronds narrow-oblong, the lobes tapering to a point. The scales of the stipes are broad, blunt, and whole-coloured, and the caudex is creeping. This connects *cristata* with *spinulosa.*

Lastrea spinulosa grows erect; has narrow, lance-shaped, bipinnate fronds, rather more deeply divided

* *Nature-Printed British Ferns,* octavo edition, 1863.

than the foregoing. The scales of the stipes are blunt
and whole-coloured, and the caudex creeps.

Lastrea dilatata spreads more, and has broader or
ovate lance-shaped fronds. The stipes is clothed with
lance-shaped scales, darker-coloured in the centre than
at the margins. The caudex is erect.

Lastrea œmula is spreading, evergreen, and has
fronds smaller than those of *dilatata*, triangular, bipinnate, the lobes having their edges curved back so as
to present a hollow upper surface. The scales are
narrow, pointed, and jagged; and the caudex is erect.

The Narrow Prickly-Toothed Buckler Fern — L.
SPINULOSA (sometimes *spinosa*) — has a stout stem, or
caudex, either decumbent or slowly creeping horizontally, with the fronds growing erect from its apex; the
fronds branched, sometimes tufted, slightly scaly,
formed of the enlarged and enduring bases of the de-
cayed fronds, surrounding a woody axis, the scales
resembling those of the stipes. The fronds are from a
foot to three or four feet high, bipinnate, the pinnæ
obliquely tapering, the inferior pinnules being larger
than the superior. This is most obvious at the base of
the frond, where the pinnæ are broader than they are
toward the apex. The lower pinnules on the basal
pinnæ are oblong, narrowing upwards, the margins
deeply cut, the lobes being serrated, and the teeth
somewhat spinulose; those toward the apex of each
pinna, as well as the basal ones of the pinnæ nearer
the apex of the frond, become gradually less and less
compound, so that, although the margins are still furnished with spinulose teeth, they gradually lose the

deep lobes which are found on the lowest pinnæ. In
all the more compound Ferns there is a similar difference of form according to the disposition of the pinnules, and in all such cases it is usual only to describe
the most complete — that is, those at the base of a few
of the lowermost pinnæ. The venation in the less
divided pinnules consists of a midvein giving off
branched venules, the sori borne on the lower anterior
venules proceeding from these, about midway between
the vein and the margin, thus forming an even double
row on each pinnule. The same arrangement occurs
on the lobes when the pinnule is more divided. The
indusia are kidney-shaped, with the margin entire.

L. *spinulosa* is common over the whole of England,
generally in moist shady places, ranging from the sea-
level to an elevation of 600 feet. In Scotland, Wales,
and Ireland it is rare. In the Lake Country it is
common in bogs and damp woods throughout the
district.

THE

BROAD PRICKLY-TOOTHED
BUCKLER FERN.

Lastrea dilatata. — PRESL.

The Broad Prickly-toothed Buckler is one of the
most compound and handsome, as well as one of the
most common, of our native Ferns, growing in broad
arched fronds, from a large tufted stem, to, when most
luxuriant, even the height of five feet, always more
or less drooping or curved. It is a species very difficult to understand, on account of its many varieties,
— some of which pass almost into *L. spinulosa* on the
one side, and others into *L. æmula* on the other. The
distinguishing characteristics, however, of *L. dilatata*
in the group of Crested Shield Ferns, of which it forms
a very large proportion, are its lance-shaped dark-

centred scales and its gland-fringed indusia. The following description applies to the more usual or typical form of *dilatata*.

Fronds ovate, lance-shaped in general outline, on a stipes of moderate length much thickened at the base and densely clothed with entire lance-shaped pointed scales very dark brown in their centres but nearly transparent at their margins ; bipinnate, with elongate-triangular, or tapering, pinnæ, placed nearly opposite, and more or less obliquely, from the larger development of the lower side. Pinnæ pinnate, pinnules near their base often so deeply divided as to be again almost pinnate, the rest pinnatifid or in the upper parts merely deeply serrated, but the margins, whether deeply or shallowly lobed, set with teeth ending in short spinous points. Venation similar to the more compound parts of the allied species. Sori in great plenty, ranged in double lines across the larger lobes of the pinnæ or along the less divided parts, and covered by kidney-shaped scales or indusia fringed round their margins with projecting glands.

The typical form of *L. dilatata* grows nearly all over the United Kingdom, from the coast-level to an elevation of 3,000 feet. It prefers shady situations, moist woods and glens, thickets and hedgerows. It is widely dispersed through the northern hemisphere, and in the Hookerian Herbarium is a specimen labelled " from New Zealand." It is common everywhere throughout the Lake District.

L. dumetorum is a distinct variety of *dilatata*. Its type may be taken from one found by Miss M. Beever, dwarfish, with broad-ovate or elongate-triangular and sometimes deltoid fronds, remarkable for their glandular surface, and for the large abundant sori produced freely on plants of a very immature age. Some of its modifications have been referred to var. *collina*, from which, however, they differ in their abundant glands and fimbriated or jagged scales. Miss Beever's plant was found on the fells of Silverthwaite, Westmorland, and the same form has been gathered by Mr. Clowes near Hawes Water, and by the Rev. G. Pinder near Elter Water.

L. collina is another distinct and permanent variety, a remarkably neat and elegant plant, growing erectly, the frond having sometimes an ovate outline alternately elongated at the apex, sometimes more elongated, oblong-lanceolate or ovate-lanceolate, dark green, a foot to two feet high, smooth or sparingly glandular, bipinnate. The stipes varies from one-half to one-third of the frond, green above, tinged with dark purply-brown at the base, with entire lanceolate dark-brown scales, conspicuously darker in the centre. The scales narrow, with a long subnate point, at the base of the stipes, where they are most numerous, broader and shorter higher up; the rachis almost without. The pinnæ, especially the lower, distant and spreading, the lowest pair unequally deltoid, the next more elongate and less unequal, the rest narrower, parallel-sided, rounding slightly near the end to an acutish point, not acuminate. Pinnules convex, obtusely oblong-ovate, the basal narrowed to a broadish stalk-like attachment, the rest sessile and more or less decurrent; the larger pinnules deeply pinnatifid, with **blunt** oblong lobes, sparingly toothed, the teeth coarse acuminately aristate (**or** bearded), mostly **at the apex. Sori mostly arranged** in two lines **along the pinnules, as in the** smaller **forms of** the species, and **covered by** glandfringed indusia. This variety was first brought into notice **by** the **Rev. G.** Pinder, found by Elter Water, in Langdale, and **by Mr.** Ecclestone **at** Torver, near Coniston. This **last** is rather larger and more divided, with concave pinnæ and strongly **convex** pinnules; it is also somewhat **glandular.**

L. alpina is remarkable for being more delicate and membranaceous than other forms of the species; the fronds normally oblong, **or** straightsided, with **the** point tapered off as in the typical *spinulosa*, but in some specimens even broadly ovate, almost **or** quite tripinnate below, bipinnate **upwards. Found by Mr.** Clowes at Hawes **Water.**

OTHER VARIETIES.

contracta, — **Mardale, J. M.** Barnes.
interrupta, — Witherslack, J. M. **Barnes**; Windermere, F. Clowes.
irregularis, — Witherslack, **J. M. Barnes.**
stenophylla, — **Witherslack, J.** M. Barnes.
glandulosa, — **Windermere,** F. Clowes; Levens, J. Crossfield.
tenera, — **Windermere, F.** Clowes.
alpina, — High Street, **F. Clowes.**
dumetorum, — Coniston, **Miss Beever;** Mardale, F. Clowes.
collina, — Coniston, Miss Beever.
grandidens, — Lindal, J. Crossfield.
Howardii, — Levens, M. Stabler.

THE HAY-SCENTED BUCKLER FERN,

TRIANGULAR PRICKLY-TOOTHED, OR CONCAVE.

Lastrea æmula. — BRACKENRIDGE.

The Hay-scented **Fern is a plant** of from a foot to
two feet in height, growing **in a** circle of triangular
arched or drooping fronds with a crisped appearance,
from the turning back of the margins of all the segments. Its fragrance is like that of new hay, **like** hay,
too, **more** powerful as it dries, and lasting for a long
time. **Its** stipes **is** of about the same length as the
leafy portion **of the frond,** clothed with jagged **pale**
brown scales. The fronds are bipinnate, the lowest
pair of pinnæ being **always longer** and larger than the
rest, and the pinnules on the inferior side of the pinnæ
always larger **than** those **on the** superior. The pinnules are oblong-ovate, the lowest often again **divided**
into a series of oblong lobes, mostly decurrent, **but**
sometimes slightly stalked, the margin cut into short
spinous-pointed teeth. The veins of the pinnules
alternately branch from **a sinuous** midvein, **and** divide
again **into two or** three **alternate** venules, **the** lowest
anterior **venule bearing a** sorus, **the** exact ramification
of the veins depending on the **degree in** which the
pinnules or lobes **are** divided. The **sori** are spread
over the whole surface, in **two** tolerably even lines

along each pinnule or lobe. Indusia small and kidney-shaped, with uneven margins fringed by small round stalkless glands. The whole frond is covered with similar glands. By these stalkless, or sessile, glands, as well as by the fewer and narrower scales of the stipes, *L. æmula* is distinguished from *L. dilatata* — whose glands are stalked. In ordinary cases, the triangular outline and hollow· crisped surface of the fronds are sufficient to distinguish the Concave Buckler Fern, which is also more decidedly evergreen, and has this further peculiarity, that the fronds decay from above downwards—not like the Broad Prickly-toothed Fern, upwards—from the base. Its range of elevation does not appear to exceed 600 feet. It prefers shady and rocky localities, and is easily cultivated.

HABITATS. — St. Bees' Head (*J. Huddart*), Broughton (*J. M. Barnes*), Coniston (*Miss Beever*), Windermere (*F. Clowes*).

THE SPLEENWORTS.

THE LADY FERN.

Athyrium Filix-fœmina.* — ROTH.

The genus *Athyrium* holds a place between the
Aspidiæ (or Shield Ferns) and the *Aspleniæ* (or
Spleenworts). Its generally elongated sori mark it,
however, as belonging rather to the latter group,
though there is a sufficient approach to the roundish
kidney shape of *Lastrea* to account for its having been
also attributed to the former. It is, nevertheless, not

---

* Polypodium Filix-fœmina (*Linnæus*), Aspidium Filix-fœmina,
Asplenium Filix-fœmina, Cystopteris Filix-fœmina, &c.

so like to *Lastrea* as to be mistaken for it, and is distinguishable also from the other *Aspleniums* by its annual fronds and its herbaceous texture.

The Lady Fern, so called because of the peculiar delicacy of its fronds contrasted with the masculine robustness of the Male Fern, grows like that in plume-circlets or coronals from the caudex, which in winter, whether close to the ground or a few inches above it, bears a tuft of incipient fronds, each rolled up separately and the mass nestling in a bed of chaff-like scales. In May or June they are developed, twenty or more being usually produced. In the summer a few more generally arise in the centre, the whole dying off in the autumn. The form of the fronds is lanceolate, more or less broad, the stipes scaly at the base and about a third of the length of the frond. The fronds are bipinnate, the pinnæ always lanceolate, more or less drawn out at the point, and always again pinnate, though sometimes with the bases of the pinnules connected by a narrow leafy wing, but not so much so as to render them merely pinnatifid. The pinnules, however, are more or less lobed or pinnatifid, the lobes being sharply toothed in a varying manner. The venation, owing to the delicate texture of the frond, is very distinct, consisting in each pinnule of a wavy midvein, with alternate and again alternate venules, on the anterior side of which, at some distance from the margin, is an oblong sorus. In the larger and more divided pinnules the venation is more compound, and more than one sorus is borne on each primary vein, which thus becomes a midvein with branches on a

smaller scale. The sori are slightly curved, the basal
very much so, being horse-shoe shaped; the indusia of
the same form. This horse-shoe shape is made by the
lateral line of spore cases crossing the vein and then
returning, and sometimes the indusium is circular all
but a small notch, so somewhat resembling the fructification of *Lastrea*. One side of the indusium is fixed
lengthwise to the side of the vein which forms the
receptacle, while the anterior one (that toward the
midvein of the pinnule) becomes free, and is split into a
fringe of hair-like segments.

The Lady Fern is common all over England and
Ireland, less so in Wales and Scotland (in the Highlands at an altitude of 3,000 feet), but found in all our
Northern, Western, and Channel Islands; it is found
also in one or other of its forms from Lapland to
Crete, from the Ural mountains to Kamtchatka, from
the Mediterranean to India, from Abyssinia to Algeria,
from Canada to British Columbia, and in the United
States and South America. It is perhaps the most
prolific in varieties of all our British species, the
varieties being very marked, singular, and permanent.
It is common everywhere in the Lake Country.

VARIETIES.

Barnesii, — **Levens, J. M.** Barnes.
brachypterum, — Whinfell, J. M. Barnes.
defectum, — Tebay, **J. M.** Barnes.
erosum, — Brigsteer, **J. M.** Barnes; Windermere, **F.** Clowes.
exiguum, — Levens, **J. M.** Barnes.
fimbriatum, — **Farleton** Knott, J. **M.** Barnes.

folioso-multifidum, — Old Hutton, A. B. Taylor.

grandissimum, — Levens, **J. M.** Barnes.

limbo-spermum, — Tebay, **J. M.** Barnes.

multicuspis, — Levens, **J. M.** Barnes.

parvicapitatum, — Witherslack, J. M. Barnes.

*plumosum **Barnesii,*** — Milnthorpe, J. M. Barnes.

rhaticum-multifidum, — Burneside, **A.** B. Taylor.

subcruciforme, — Whitbarrow, **J. M. Barnes.**

subdigitatum, — Burneside, **Martindale,** A. B. Taylor.

uncum, — Arnside, J. **M. Barnes.**

laciniatum, — Newby **Bridge, Wollaston.**

marinum, — **Windermere, J. Wood.**

oxydens, — **Windermere, C. Monkman.**

curtum, — **Windermere, J. Wood;** Levens, **J. M.** Barnes.

ramulosum, — **Windermere, F. Clowes.**

subdepauperatum, — Windermere, **F. Clowes.**

flexuosum, — Silverdale, J. Crossfield; Windermere, J. Huddart.

latifolium, — Keswick, Miss Wright; Arnside, **J.** Crossfield.

Monkmani, — Troutbeck, C. Monkman.

strigosum, — Burton, **J. Jones.**

exile, — Levens, J. **M. Barnes;** Burton, **J.** Jones.

polydactylum, — **Dent, A.** Mason; Windermere, F. Clowes.

crispatum, — Arnside, **J.** Crossfield.

laciniatum-dissectum, — Levens, **J. M.** Barnes.

BLACK MAIDEN-HAIR SPLEENWORT.

Asplenium Adiantum-nigrum. — LINNÆUS.

The true Spleenworts (so called from some old-time supposed virtue in curing diseases of the spleen) are small evergreens, known from all other of our native Ferns, except *Ceterach,* by the long narrow single sori lying in the direction of the veins which traverse the fronds,— *Ceterach* being distinguished from them by having the backs of its fronds clothed with brown scales, under which the sori are hidden. From their

F

next neighbours, the *Athyriums*, they are known by the latter **having** hippocrepiform (or horse-shoe shaped) **sori and the free margins of the indusia fringed,** while in the *Aspleniums* the **sori are** not curved, and the margins of the indusia are but slightly jagged, **if not quite entire.** The Spleenworts, too, **are evergreen ;** while **the Lady Fern is** deciduous. **There are** nine British **species of** ASPLENIUM — *fontanum, lanceolatum, Adiantum-nigrum,* **marinum,** *Trichomanes,* **viride,** *Ruta-muraria,* **germanicum, and** *septentrionale.* **Of these,** *A. fontanum* **and** *A. lanceolatum* do not belong to **the** Lake District, **though** there is a tradition of the former having inhabited **Wythburn** (found there by Hudson, about 1775) until exterminated by the "greed **of** collectors." Let the race take heed !

The Black Maiden-hair Spleenwort is **an evergreen,** growing in tufts, **and varying in** height **from three or four inches to eighteen or more, including** the stipes, which **is often as** long as **or longer than** the leafy portion, **except in stunted** specimens. The stipes **is of a** shining dark purple. The fronds are either erect or drooping, according to situation, **of** a thick leathery **texture, triangular,** more or less **elongated** toward the point, bipinnate, sometimes tripinnate ; the pinnæ pinnate, triangular-ovate and elongated at the point, the lower pair longer than the next above them ; the pinnules, especially on the larger pinnæ, again pinnate, the alternate pinnules deeply **lobed** and the margins sharply serrated. The veins are numerous, each pinnule having its distinct midvein, branching **into** simple **or** farther-branching veins, on which the sori are produced near the junction with the midvein, — that is to say, near the centre of every lobe or pinnule. All the ultimate divisions of the fronds, as well as all the larger lobes, have midveins with these simple or branched venules. When young, the sori are distinct and of the elongated narrow form common to the genus, but, growing older, they spread till they often become confluent and cover the entire under-surface of the frond. The indusium is narrow, its margin free and entire ; but it is soon lost, being pushed away by the growing sori. This species is very variable : in dry and exposed places small and obtuse, in more sheltered drawn out and elongated. These extreme states are ranked as varieties. Some have been found also with the fronds variegated with white. It seems to be nearly as common as the *Athyriums*, though not growing to so great an elevation, for it is found nearly everywhere, from Scandinavia to the Cape of Good Hope, in the Sandwich Islands, in Affghanistan, in Java, and in St. Helena. It thrives moderately well in cultivation if planted in a sandy soil and well drained ; and is easily manageable as a pot-plant, but requires a pure atmosphere. It is common throughout the Lake Country, either on the slate or limestone, often preferring old walls.

VARIETIES.

intermedium, — Heversham, J. M. Barnes.
depauperatum, — Windermere, F. Clowes.
acuti-dentatum, — Witherslack, J. Crossfield.

THE SEA SPLEENWORT.

Asplenium marinum. — LINNÆUS.

A sea-side Fern, as its name denotes, but occasionally found inland; a tufted evergreen species, erect or decumbent, the **fronds** growing usually six or eight **inches** long, **linear or linear lanceolate,** of the deepest **glossy** green, **and of a** leathery texture, with **a stipes** shorter **than** the **frond,** smooth, channeled **in front,** chestnut-coloured **or** purply-brown. The fronds **are** simply pinnate, with stalked pinnæ, connected by a narrow wing extending along the rachis; obtusely ovate or oblong, unequal at the base, the anterior base being much developed, while the posterior appears cut away, with the margin serrated or crenated. The venation is tolerably distinct : each pinna has a mid-vein, giving off veins alternately on either side, branching again into a series of venules. The sori, lying obliquely on the anterior side of each venule, form two rows on each side of the centre ; oblong or linear, with persistent indusia opening along the anterior margin as the spores ripen. The chief variation of the Sea Spleenwort consists in the elongation of its parts, the pinnæ sometimes tapering to a narrow point, sometimes also being auricled at the base and deeply lobed. It keeps very close to the sea-level. In cultivation it requires warmth, and grows best in sandy peat-soil, in the interstices of stones or rock-work.

The Sea Spleenwort is most abundant on the west coast of Europe, extending, however, eastward in the Mediterranean. It is found also on the African coast, in the Western Isles, and, according to Sir W. Hooker in St. Helena. This peculiar distribution has been supposed to indicate that it took place prior to the great disruption of the chalk and the vast deposit of allurial mattar along the eastern coast of England.

HABITATS. — Whitehaven, St. Bees' Head, Head of Morecambe Bay, Sea-cave near Silverdale, Piel Castle, Heysham.

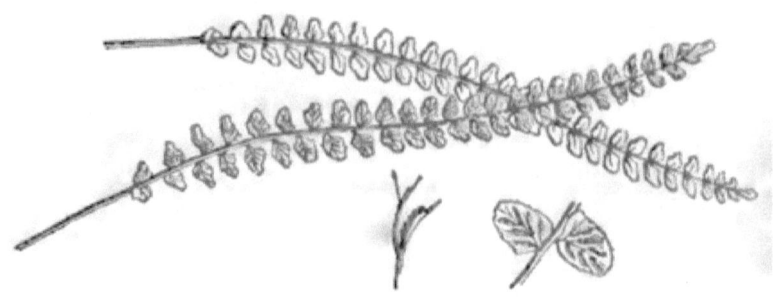

COMMON MAIDEN-HAIR SPLEENWORT.

Asplenium Trichomanes. — **LINNÆUS.**

The Common Maiden-hair Spleenwort is **but a diminutive plant**, yet it is one of the most elegant **of the**
hardy **evergreens**, noticeable **for the** contrast between
its purply-black stipes **(and** rachis) **and bright green
pinnæ**, and for the regularity with **which** the latter are
disposed. Its numerous small slender fronds, generally
not more **than** from three to six inches long, though
sometimes **double** that, grow **in** tufts in rock crannies,
and delight **in** the crevices of old walls. Its fronds
are simply pinnate, the **pinnæ** small and numerous,
equal-sized, roundishly-oblong, attached to the rachis
by a stalk-like projection **of** their posterior base, **the**
margins **entire or crenated (with convex or round**
teeth). The pinnæ are jointed to the rachis, and when
old are readily displaced, leaving the black naked
rachis among the other fronds. **A distinct** midvein
passes through **each pinna, branching on each** side into

veins and venules, the anterior bearing the linear sori just within the margin of the pinnæ. The sori, when young, have a thin indusium, with a rather round-toothed free margin, but when older become confluent, and cover the whole frond. This also is a very universal Fern, found not only in the Old and New World, but also in the newer world of Australia. It once had a medicinal reputation, and, according to Ray, was usefully employed in affections of the chest. It is also sometimes referred to in old medical books as the plant from which the syrup called Capillaire is produced. Turner, in his *Herball*, published in 1568 calls it "English Mayden's Heare," and says:—"the juice stayeth the heare that falleth of, and if they be fallen off, it restoreth them agayne." It grows best in cultivation, in sandy loam with leaf-mould, and does not require so much shade as other ferns.

HABITATS.—Ambleside, Keswick, Borrowdale, and Calder Bridge.

The most beautiful of its VARIETIES is the *incisum*, with pinnæ deeply pinnatifid with linear notched segments. It has been found in Borrowdale by Miss Wright, and in Lindale-in-Cartmel by Mr. A. Mason.

OTHER VARIETIES.

depauperatum,—Kendal Fell, J. M. Barnes; Whitbarrow, F. Clowes; Arnside, J. Crossfield.

bifurcum,—Windermere, F. Clowes; Arnside, J. Crossfield.

multifidum,—Windermere, F. Clowes; Keswick, Miss Wright; Ulverston, Mrs. Hodson.

ramosum,—Arnside, J. Crossfield.

rotundato-multifidum,—Witherslack, J. M. Barnes.

THE GREEN SPLEENWORT.

Asplenium viride. — HUDSON.

The Green Spleenwort has so close a resemblance to
the Common Maiden-hair as to be often mistaken for
it by hasty observers. It is distinguished by its green
compressed rachis (that of *A. Trichomanes* being dark
brown or black), by its persistent pinnæ (deciduous in
A. Trichomanes), by the more central situation of its
sori, which are placed rather below than above the
vein-fork, and by being always of a much paler green
and of a more delicate herbaceous appearance. It is
an evergreen tufted species, with bright pale green
fronds, narrow, linear, simply pinnate, from two to
eight inches long, supported by a short stipes, dark at
the very base, but else green, the rachis all green.
The pinnæ are small, generally roundish-ovate, slightly
taper toward the base, and attached to the rachis by
the narrowed stalk-like part, the margin being deeply
crenated. The venation is distinct ; the midvein
sends off alternately a series of venules, either simple
or forked, which have the sori on their anterior side.
The sori are oblong, covered at first by membranous
indusia, which are soon pushed aside ; the free margin
is jagged or crenate.

* Asplenium **Trichomanes** ramosum (*Linnæus*), Asplenium intermedium (*Presl*).

The **Green** Spleenwort **is found in** most moist, **rocky,** mountainous districts **of** Great Britain ; **it oc-curs** also, though **less** frequently, in **Ireland, and** throughout Europe. It is to **be** cultivated in pots in a close damp **frame ; or** on moist shady rock-work, **if** covered with **a** bell-glass. If exposed, it **is apt to** suffer **from** occasional excessive wet, which **often does** not properly **drain away ;** and also from **the dry hot** air of **summer.** The glass protects **it** from both these casualties, and provided it is not kept too close it will **then** thrive well. The proper bell-glasses for these half-hardy Ferns have a small opening in the crown, **which** may be closed or not at pleasure, but which is generally best left open. In **pots, the** plants should have a gritty, porous soil.

HABITATS. — Ambleside, **Patterdale,** Kendal **Fell** (*W. Christy*), Hutton Roof, Farleton, Arnside (*Rev. G. Pinder*), Casterton Fell, Mazebeck Scar (*R. B. Bowman*), Borrowdale, Ashness Ghyll, Barrow Force, Gillsland, Brandy Ghyll on Carrock Fell, Whitbarrow, &c.

VARIETIES.

multifidum, — Farleton Knott, **J. Jones ;** Scout Scar, **J. Crossfield,** J. Wood.

subbipinnatum, — Whitbarrow, J. Huddart.

varians, — Kendal Fell, **J. M. Barnes.**

THE RUE-LEAVED SPLEENWORT,

OR WALL RUE.

Asplenium Ruta-muraria.— LINNÆUS.

This is a very diminutive Fern, growing, as its name implies, upon old walls, and very common on the limestone rocks, like the Rue in general appearance; sometimes not above an inch high, seldom in the most favourable situations reaching to the height of six inches. Its fronds are numerous, of bloom-covered (glaucous) green, usually triangular in outline, bipinnate, and with a stipes about half the entire length of the plant. The pinnæ are alternate, with rhomboidal, or roundish-ovate, or obovate pinnules, the base wedge-shaped, tapering into a more or less distinct petiole, the apex rounded or truncate, or sometimes acutely

 prolonged, always toothed with small or nearly equal teeth. The more luxuriant fronds become almost tri-pinnate, the pinnules deeply pinnatifid, and the lobes formed like the ordinary pinnules. When the plants are quite young, the fronds are simple and roundish kidney-shaped. At a later stage they are occasionally only once pinnate, with pinnatifid pinnæ. The upper margins of the pinnules are irreg-

ularly-toothed. The venation consists of a series of veins repeatedly forked from the base, so that there is no distinct midvein; the number of the venules corresponds with the number of marginal teeth. Several sori are produced near the centre of the pinna, covered by indusia opening inwardly with a jagged or irregularly-sinuated margin. The plant is evergreen and easy of cultivation. It is so common that there is no occasion to give any special habitats. It extends to about 600 feet above the sea-line.

In its normal conditions *A. Ruta-muraria* is easily recognizable: the characters afforded by its triangular (deltoid) outline, bipinnate or tripinnate division, and distinct wedge-shaped pinnules, together with the smallness of the fronds, sufficiently distinguishing it from the other *Aspleniums*. There are, however, certain of its forms which are not, at first sight, so easily separated from *A. germanicum*, being narrow on the fronds or pinnules, and sometimes scarcely more than pinnate. These forms are best distinguished by the round-toothed (crenulate) indusia, and by the fine even toothing of the anterior margin, — the indusium in *A. germanicum* being entire, and the apex of its pinnules being less deeply and unequally notched.

VARIETIES.

cristatum, — **Farleton** Knott, Wollaston.
unilaterale, — Troutbeck, Miss Wright; Kendal Fell, J. M. Barnes.
cuneatum, — Sizergh, **J.** Crossfield.
ramosum, — Silverdale, J. Crossfield.

ALTERNATE-LEAVED SPLEENWORT.

Asplenium germanicum. — WEIS.

The Alternate-leaved Spleenwort stands between the
Wall Rue and **the** Forked Spleenwort, sometimes, indeed, marked as a dubious species, but decided by
Moore to be perfectly distinct. It is **one** of the rarest
of our native Ferns, rare also in Northern and Central
Europe. **In** other parts of the world it is not known.
It **is so** rare here in Great Britain that Moore **records**
only one single variety. **Its** altitudinal range **is from**
300 to 600 feet above the sea.

The Alternate Spleenwort grows **in** tufts, the fronds
from three to six inches high, sub-evergreen (the fronds
more or less persistent), **narrow** linear in general outline, pinnate, **divided** into distinct, alternate, wedge-
shaped pinnæ, **one or** two of the lowest having generally a **pair of very** deeply-divided lobes, **the** upper
more and **more** slightly **lobed,** all having their upper
ends **toothed or** notched. **The venation is** very indistinct, **on** account **both of the** narrowness of the **parts**
of so small fronds **and of** their opacity. **There is** no
midvein, but one of the venules extends to **each of** the
teeth, each vein entering from **the** base becoming twice
or thrice branched **as** it reaches the broader parts upwards, six or eight veins generally lying near together

in a narrow fan-like manner in each of the larger pinnæ, the smaller having proportionably fewer. Two or three linear sori are produced on a pinna, covered by membranous indusia, the free margin of which is entire, or slightly sinuous, but not jagged. The sori at length become confluent. It is very difficult of cultivation.

For the cultivation of *A. germanicum,* Moore (our chief authority) recommends that it should be potted in sandy peat-soil, well drained by a mixture of rubbly matter (indeed, good drainage seems indispensable to almost all of the Fern kind); and that it should be kept under a bell-glass in a shaded frame or greenhouse. The plants are very liable to die in winter, the best safeguard from which is not to allow any water to lodge about the crowns, nor to keep the bell-glass too closely or too constantly over them.

HABITATS. — Borrowdale (*Miss Wright* and *H. E. Smith*), and near Scawfell (*Rev. H. W. Hawker* in an excursion with *J. Huddart* and *F. Clowes.*

THE FORKED SPLEENWORT.

Asplenium septentrionale. — HULL.

This is another of the small and rare Ferns, though more widely distributed than *A. germanicum*, and growing to an elevation of 3,000 feet, tufted sometimes in large masses and grassy looking, differing from *A. germanicum* (which some botanists consider a variety of it) by its fronds being either simple with mere lobes, or forked with two distinct branches, each like its own smaller fronds, and never being regularly pinnate as *A. germanicum* is. It is also narrower in its parts, with a thicker texture, and less leafy. The fronds are from two to six inches long, slender, and of a dull green; the stipes is rather long and dark purple at the base; the leafy part of the frond, hardly to be called leafy, is narrow elongated lance-shaped, split near the end into two or sometimes three alternate divisions, or in the smaller fronds into as many teeth, each of the divisions of the frond having its margin cut into two or more sharp-pointed teeth, the points of the larger teeth very frequently split again. The forked fronds are indefinite in form and apparently one-sided, one division being smaller than the other, and looking like a side branch with nothing to balance

it on the other side of the rachis. The lobes are sometimes so much separated as to look like distinct pinnæ.
There is no midrib or vein, the rachis answering the
purpose if the frond is not lobed, or else becoming
forked so as to send up one vein to each of the teeth.
Three or four long linear sori are crowed into this
small space, so that when the ripening sori burst the
indusia, they become confluent over the whole under-
surface. This confluence of the sori over the whole
under-surface has led some writers to consider this
plant an *Acrostichum*. Others, from the sori being
face to face in consequence of their growing on each
side of the vein and almost close, have thought it a
Scolopendrium, the mark of which is to have the sori
confluent in pairs face to face. It has therefore been
sometimes called *Acrostichum septentrionale* and *Scolopendrium septentrionale*. If the plant, however, be
examined when young, it will be found to be a true
Asplenium.

The Forked Spleenwort does not appear to be found
in Ireland; but, though rare, has a wide range in
Great Britain, from Devonshire to the Orkneys. It
grows abundantly in some of the mountainous tracts
of Central Europe, and extends from Russia and Scandinavia to Italy and Spain. In Asia it inhabits the
mountain ranges of the Ural and the Altai, and is
found from Northern India to the Caucasus. It
occurs also in New Mexico. It prefers fissures of
rocks, or between the stones of loose walls.

As in the case of the allied species (Moore again, in
the octavo edition of his *Nature-printed British Ferns*),

many persons fail to cultivate this **Fern** with success ; probably from the use of fine soil **in too large** masses. Naturally this is **a** rupestral (rock-growing) plant, and **this** condition should be imitated by **its being planted** among masses **of** porous sand-stone in the interstices **of which, and only** in the interstices, a little sandy **soil should be** placed. It would **no** doubt also **be an advantage** to plant somewhat horizontally instead **of too vertically and to have the upper stones large** enough to shade **the crowns of the plants from the sun.** Many **Ferns do not** need **so much shade as is** given in a collection **to** the tenderer **sorts ; and this partial shadow-** ing would be **more** congenial to **some of the wall or** rock **species than** a more general exclusion **of the sun.**

HABITATS.—Scawfell, by Wastwater, Honister Crags, Borrowdale, Newlands, Keswick, Helvellyn, Patterdale, Red Screes (Ambleside), Crummock Water.

THE
COMMON
HART'S TONGUE
FERN.

Scolopendrium vulgare.* — SMITH.

Scolopendrium forms a sub-group of the *Asplenieæ*, in which the sori, instead of being simple and distinct, are brought together in opposite pairs, so that what appears to be a line of spore-cases forming a sorus is in fact a double line forming a double or twin sorus. The name *Scolopendrium* is from *Scolopendra*, a centipede, from some supposed resemblance between the

---

* Asplenium scolopendrium (*Linnæus*), Blechnum linguifolium, &c.

G

feet of the centipede and the lines of fructification of
the Fern. Its English name comes from the likeness
of the whole frond to the shape of a Hart's tongue,
differing altogether from the generality of its feathery
fellows, in being only one long shining bright green
leaf, partially erect when dwarf, but drooping in its
larger development, growing in groups or tufts, on
rocks, damp masonry, and moist banks, from four
inches to upwards of two feet long, hanging down indeed like great tongues lolling over the grey walls or
grassy banks.

The dwarf fronds are thick and of a leathery texture,
the larger thinner and less rigid; their outline is what
is called strap-shaped, that is narrow oblong lanceolate
much elongated; they taper toward a point at the
apex, and become narrower downwards, cordate
(rounded like the thick end of the heart in cards) at
the base; the margin is entire, or very slightly wavy;
and the stipes is shaggy and about half the length of
the leaf. The venation consists of a strong midvein or
midrib, extending the whole length of the frond, from
which run forked veins, their branches parallel and
proceeding direct toward the margin, terminating just
within it in a club-shaped apex. The veins are usually
forked twice, but are not constant to any exact number of divisions. The sori, which are oblong patches
of unequal length, lying in the direction of the veins
at short intervals along the upper two-thirds of the
frond, are composed each of two lines of fructification
united at their sides, each of these lines, however,
consisting of a complete sorus, so that the two united

are properly called a twin sorus. This **twin sorus is** always produced between two fascicles of veins: that is, the lowermost venule produced by **one vein** and **the** uppermost of another below become each **a** receptacle upon which one of the two contiguous lines **of** sporecases is produced. Their indusia are attached on **the** upper **and** lower sides of the venules, the other edges overlapping, **so that the free margin is outside the fas**-cicle **of** venules to which it belongs. **When very** young, the separation where **they overlap is not ap**-parent, **but** becomes **so as they advance** toward **matu**-rity; till at last they open down the centre, one **in**-dusium turning up and the other down, the **two lines** of spore-cases becoming confluent and confused.

This is the normal condition **of** *Scolopendrium*; **but** the genus **is** one of the most remarkably prolific in varieties, and in remarkable **varieties**, among all known Ferns, the greater part of these, though not **un**-frequently altogether monstrous, permanent and **re**-taining their peculiarities in cultivation. **Several forms** very distinct in themselves, and distinct **also** from **the** parent, **have** been produced from the spores by artificial treatment, indicating how **probably** the same process is going on **in** a state of nature, more slowly perhaps and imperceptibly, **but as** certainly, giving rise to **new** forms, some transient, but **some to be** perpetuated.

The fructification of *Scolopendrium* is, of course, as being one of the *Polypodiaceæ*, normally dorsal; but in some of its varieties there **is** a **very curious deviation** from the law : the sori **are produced on the upper as** well as the lower surface, and **sometimes abundantly** so. This occasionally happens from the elongation of the normally-placed sorus of the underside, which extends to the margin and returns on the upper side when the sori are placed opposite the marginal crenatures. But it also frequently happens that the sori are produced on the upper side distinctly within the margin, and where there are no corresponding sori beneath. Those varieties which have the margins toothed (crenated) or lobed seem most liable to assume this suprasoriferous (bearing-the-sori-on-the-upper-surface) condition.

The Hart's Tongue seems to be pretty generally distributed throughout England, Scotland (more sparingly), Ireland and the Channel Islands, and through most countries in the northern temperate zone. It is not a sea Fern, but prefers the coast neighbourhood in more northern latitudes, ranging to an elevation of some 500 or 600 feet. With all its varieties, it is hardly to be considered a common Fern. It may be easily cultivated, and no Fernery should be without some few at least of its endless changes. Medicinal virtues have been attributed to it; among others, that of being good for burns and scalds. It is abundant in the limestone districts, and found also here and there upon the slate.

VARIETIES.

bimarginatum, — Ulverston, Mr. Hadwin ; Beetham, J. Crossfield; Whitbarrow, J. M. Barnes.

contractum, — Levens, J. M. Barnes.

crispum-soriferum, — Whitbarrow, A. B. Taylor.

crista galli,— Levens, **J. M. Barnes**; Whitbarrow, F. Clowes; Arnside, J. M. Barnes.

cristatum, — Whitbarrow, **J. M. Barnes**; Beetham, J. Crossfield.

cymbiformæ, — Whitbarrow, J. M. Barnes, G. Stable.

fissum, — Brigsteer, **J. M. Barnes**; Silverdale, J. Crossfield.

glomeratum, — Storth, Miss Nicholson.

incisum,— Levens, J. M. Barnes.

intermuricatum, — Milnthorpe, J. M. Barnes.

laciniatum, — Whitbarrow, **J. M. Barnes.**

laciniato-marginatum, — Levens, **J. M. Barnes.**

lineare, — Whitbarrow, J. M. Barnes.

marginatum, — Whitbarrow and elsewhere, F. Clowes, J. M. Barnes, J. Crossfield.

muricatum, — Farleton Knott, J. Jones.

polycuspis-transversum, — Brigsteer, J. M. Barnes.

———— *undosum*, — Levens, J. M. Barnes; Beetham, J. Crossfield.

polymorphum, — Levens, J. M. Barnes.

projectum, — Milnthorpe, J. M. Barnes; Arnside, J. Crossfield.

reniforme, — Levens, J. M. Barnes.

rimosum, — Levens, J. M. Barnes.

rugoso-marginatum, — Levens, J. M. Barnes.

sagittato-crispum, — Milnthorpe, Miss Wilson.

————*nudisorum*, — Milnthorpe, J. M. Barnes.

supralineatum, — Levens, J. M. Barnes.

supralineato-fimbriatum,— Levens, J. M. Barnes; near Ulverston, Mrs. Hodgson.

undulosum, — Levens, Garnett.

variegatum, — Arnside, J. Crossfield.

imperfectum, — Whitbarrow, Wollaston.

ramosum, — Grange, Dawthwaite.

variabile, — Heversham, J. M. Barnes; near Ulverston, Mrs. Hodgson.

sculpturatum, — Farleton Knott, J. Jones; Ulverston, Miss A. Hodgson.

macrosorum,—Wharton, Miss A. Hodgson; Arnside, J. Crossfield.

crispatum, — Furness Abbey, Mrs. Hodgson.

exsertum, — Grange, Miss A. Hodgson.

pocelliferum, — Farleton Knott.

cornutum, — Silverdale, J. Crossfield.

crispum, — Ulverston, J. Crossfield.

digitatum, — Cark, **J.** Crossfield.

divaricutum, — Arnside, **J.** Crossfield.

limbospermum, — Beetham Fell, J. Crossfield.

polyschides, — Arnside, J. Crossfield; Holmes, J. M. Barnes.

cornuto-abruptum, — J. Crossfield.

albescens, — Arnside, J. Crossfield.

THE COMMON SCALE FERN,

or Scaly Spleenwort.

*Asplenium Ceterach.** — Linnæus.

Ceterach, the botanical name of this genus (of which there is only one British species), is said to be a corruption of *Chetherak*, the name given to it by Arabian or Persian medical writers. Its old English designation of Milt-wast is said also to be a corruption : the *Milt* being the *Spleen*, and *wast* said to be from *waste*, because of some story of its destroying the spleen, — but more probably, as Bailey puts it in his good old

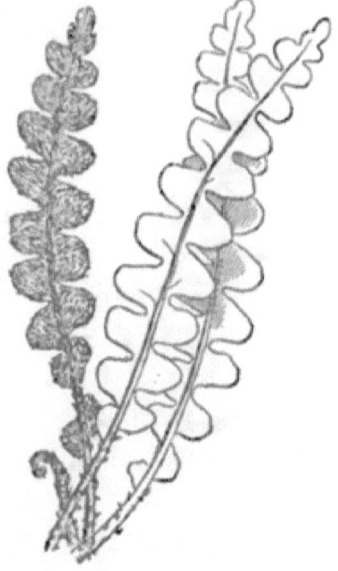

dictionary, " *Milt-wast, wort, Herbs* " (making *wast* the synonym of *wort*, a herb), *Milt-wast* is simply *Spleen-wort*, and no corruption at all. It was also called Finger Ferne, " because," says Turner in his *Herbal* (1551), "it is no longer than a manne's finger," and Scale Ferne, " because it is all full of scales in the inner syde. " The Scaly Milt-wast or Spleen-wort, growing generally about the size of "a manne's finger," sometimes not

---

* Ceterach officinarum (*Willdenow*), Scolopendrium Ceterach, Gymnogramma Ceterach, Blechnum squamosum, &c.

so large, but sometimes even six or eight inches
long, is a tufted evergreen, living on the limestone,
and lodging, when away from its native rocks, on any
old walls or ruins. The stipes is short and scaly; the
fronds are commonly pinnatifid, sometimes pinnate,
divided rather more deeply. The upper surface is a
deep opaque green; the under is densely crowded with
closely-packed and overlapping scales, whose rusty
brownness, as they project beyond the margin, seen
yet more fully in the exposed under-surface of the
young partially-developed fronds, contrasts with the
deep green of the upper surface. The pinnæ or lobes
are ovate, either entire or lobed in the margin. The
venation is indistinct, on account of the opacity of the
thick and fleshy fronds. Indeed, it is only to be made
out by examining young fronds, removing the scaly-
covering, and the outer skin of the frond itself. It is
then seen that the principal vein, entering at the lower
corner, proceeds sinuously toward the upper side of the
apex, branching alternately, and branching again, the
venules becoming more or less joined near the margin.
The sori are borne irregularly along the sides of the
venules, most of them directed toward the apex of the
pinna. At first they are quite hidden by the scales,
but ultimately the spore-cases protrude between them,
though, being nearly of the same colour, never very
obviously.

In old times this plant had a great medicinal reputation. Gerard writes of it:—"There be empiricks or
blinde practitioners of this age who teach that with
this herbe not only the hardness and swelling of the

spleene, but all infirmities of the liver, may be effectually and in a very short time removed.   .   .   .   .
But this is to be reckoned amongst the old wives' fables, and that also which Dioscorides telleth of, touching the gathering of spleenewort in the night, and other most vaine things which are found here and there scattered in old books." There may be yet some grain of truth in even old wives' fables; and *Ceterach*, though its Arabic name be lost, is still retained in Italy in the list of officinal plants. On the Welsh coast they use it as a bait in fishing for rock-cod. It does not apparently extend further north than Scotland, but is spread over the centre and south of Europe, North and "South Africa," through Central Asia, and, it is said, in Brazil. It grows freely in the garden — in lime-rubbish, requiring to be kept rather dry.

HABITATS. — Arnside Knott (*Miss Beever*), Milnthorpe, Scout's Scar (Kendal), Ambleside (*Miss S. Beever*), Gosforth (*Robson*), Keswick, Sandwith, St. Bees, Yew Crags and Ara Beck (Ullswater), Whitbarrow, &c.

VARIETIES.

ramosum, — Arnside, J. Crossfield.
crenatum, — In several places.

THE COMMON HARD FERN.

*Blechnum Spicant.** — ROTH.

Blechnum (a Latinized form of the Greek *Blechnon*) is only a Fern — any kind; but *Spicant*, erect and spike-like, as an ear of corn, well expresses the peculiar appearance of this plant, with its erect fertile frond standing above the less erected barren fronds around it. The Hard Fern, too, is a no less expressive title, for the plant is hard, rigid and hardy too. It is one of the few English Ferns producing distinct-looking kinds of fronds — fertile and barren. The barren grow in tufts, very gracefully and droopingly disposed, from six to twelve inches high; and in the centre of them, always higher, and sometimes twice their height, rises the hard upright fertile frond. The barren fronds are attached to the caudex by a very short scaly stipes, the stipes of the fertile frond also scaly; the scales long-pointed and sparse, are half the length of the whole frond, and of a dark brown. Both kinds are narrow lanceolate, the barren being only deeply pinnatifid, while the fertile are pinnate; the segments in both are long and narrow, like the teeth of a comb. The vena-

* Osmunda Spicant (*Linnæus*), Asplenium Spicant, Lomaria Spicant, Blechnum boreale, &c.

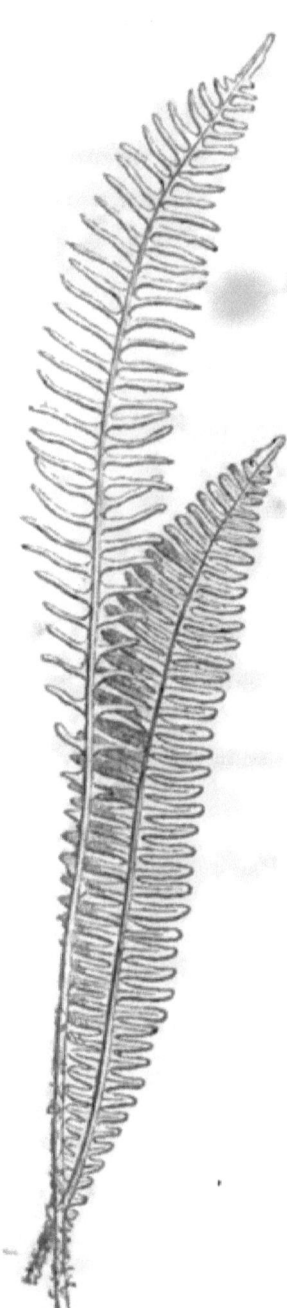

tion of the barren fronds is distinct, a stout midrib or vein producing lateral veins once or twice forked, the venules extending parallel toward the margin, and terminating in a small club-shaped head. The venation of the fertile frond, not so distinct on account of the contraction of the parts, differs in having a longitudinal venule on each side the midvein, forming the receptacle to which the spore-cases are attached. The spore-cases are arranged in two linear sori, one on each side of the midvein, distinct while young, but often becoming confluent and covering all the under-surface. The indusia, when mature, burst toward the midrib, and become split, here and there, at points opposite some of the venules. The Hard Fern is an evergreen, not large, but strong and very distinct - looking ; growing in heaths and rough stony places, in woods and shady bottoms, preferring moisture, but careless of situation, and growing in Cumberland at an elevation of 3,000 feet, in Scotland a thousand feet

above that.   From Lapland to the Cape of Good Hope,
from Japan to the Azores, from Chili to Brazil, it
grows everywhere, and is one of our most common
Ferns, yet a very elegant plant, not by any means to
be despised because it is hardy and easy of cultivation.
Coniston, Ambleside, and Ullswater are named as special places of its inhabiting in this Lake Country.

VARIETIES.

anomalum, — Witherslack, J. M. Barnes ;   in Eskdale, Isaac
 Huddart and F. Clowes.

condensum, — Witherslack, J. M. Barnes.

heterophyllum, — Kentmere, J. M. Barnes.

lineare, — Witherslack, J. M. Barnes.

longidactylum, — Brigsteer, A. B. Taylor.

polydactylon, — Sleddale Fell, J. M. Barnes.

projectum, — Borrowdale, J. M. Barnes ; Kirkby Moor, J. Crossfield.

strictum, — Coniston, Miss Beever ; Long Sleddale, J. M. Barnes ;
 Windermere, F. Clowes and J. Crossfield.

multifidum, — Windermere, F. Clowes.

dentigerum, — Windermere, F. Clowes ; Black Combe, J. M.
 Barnes.

cristatum, — Windermere, Isaac Huddart.

vellum, — Burton, J. Jones.

BRAKE, OR BRACKEN.

Pteris aquilina. — LINNÆUS.

Pteris is the Greek *pteros*, a feather, applied of old
to some kind of Fern, and well applied here, — *Pteris
aquilina*, the eagle feather, being doubly applicable to
the magnificent, however common, Bracken. *Bracken*
is Saxon ; it is the Female Fern of old writers before
Linnæus, — not to be confounded with *Thelypteris*,
the Feminine Fern, nor with *Filix-fœmina*, the Lady

Fern. *Fern* itself is old Saxon also. The Bracken
grows everywhere, except on chalk (possibly not getting
depth there), and is the commonest of all our Ferns.
Over sandy wastes, on hedge banks, in warm moist
lanes and woods, it grows abundantly, overtopping the
rankest flowers, climbing among the bushes, half supported by them, to a height of from a couple of feet to
sometimes eight or ten. The caudex, thick and blackish, is usually creeping, creeping just beneath the surface more extensively than that of any other Fern ;
but in some cases growing straight downwards to a
great depth, Mr. Newman stating that he has found it
even fifteen feet below the ground. The fronds appear
so soon as the frosts are over, coming up in little curls
like shepherds' crooks, or croziers ; sometimes like
little grey-green downy hooks stuck into the grass, the
upper part of the stipes not yet having burst the surface. The young stipes is downy and soft, growing
angular and hard in age, spindle-shaped at the base.
The fronds, erroneously said sometimes to be three-
branched, are truly bipinnate, or tripinnate when very
luxuriant, the pinnæ standing opposite in pairs, each
pair in succession becoming fully developed while the
main rachis is extending upward and the next pair is
beginning to unfold. It is only when the plant is
very poor that the fronds appear three-branched, the
development of the lower pair of branches not leaving
the plant energy enough to carry up its rachis and
produce the other pairs of pinnæ which it would normally possess. The true habit of the plant is still more
clearly shown when it attains its fullest luxuriance, the

full-grown fronds then consisting merely of a series of
pairs of branches from bottom to top. The unrolling
of the young fronds is very curious, and well worthy of
watchful notice.

The bipinnate branches, or pinnæ, are in general
ovate slightly elongated, their pinnæ (the secondary
pinnæ) narrowly lanceolate. These last are placed
rather closely together, and again divided into a series
of pinnules, which are either undivided and attached
to the rachis by their stalkless base with a line of
spore-cases along each margin, or become larger and
then more elongated and deeply pinnatifid with the
lines of spore-cases on the margins of the lobes. The
apexes of the primary and secondary pinnæ and the
pinnatifid pinnules become less and less divided, until
at last they end in a single lobe more or less elongate.
The venation is very various, depending on these
differences of development. Each pinnule has a distinct midvein, producing alternate lateral veins, which
become twice forked and extend to the margin, where
they meet a longitudinal marginal vein, which forms
the receptacle. The indusium consists of a bleached
membranous, fringed expansion of the upper skin of
the fronds, which turns back so as to cover the spore-
cases ; but there is here another membrane lying
under the spore-cases, no doubt a similar expansion of
the skin of the under-surface. The fronds are annual,
but owing to their rigid texture do not easily die off
altogether, only losing their summer verdure, standing
often through the winter, or until bowed by the
weight of snow, in all their summer glory of form, and

as gloriously beautiful in their varieties of brown as they were in their living greenness.

The Bracken grows everywhere : not only throughout our own islands, but in all parts of the world, from Lapland (at about 67 degrees north) to the Cape of Good Hope. It rises above the coast-level in the Scottish Islands to an elevation of 2,000 feet. It is useful for very many purposes. In our north country the dried fronds make capital litter for cattle ; they are also an excellent elastic material for packing and storing fruit in, a fine covering to preserve plants from frost, and make good thatch, employing the stems also. They are not bad fuel, though light and quick-burning ; and, cut green, are good manure for land, one-third of their bulk, according to Sprengel, consisting of mineral substances, potash, silica, lime, soda, chlorine, magnesia, oxide of iron, phosphoric acid, &c. The dry herbage is said to be rich in nitrogen. They are especially good for manuring potatoes. Good also for feeding pigs, who are fond of the "roots" (the underground caudices), which are succulent and starchy, and who have no objection to a jelly made by boiling the young and tender fronds. Mr. Lees suggests that the same, not made into jelly, but boiled as greens, would not be bad eaten with the pig instead of by him ; and Dr. Clarke recommends them when very young, tender, and blanched, as a substitute for asparagus. The New Zealanders eat the " roots " of a variety of the Bracken, *P. esculenta*, pounded between stones and roasted ; in Siberia these same stems are used in brewing a kind of beer, one-third fern-root to two-thirds malt ; and the

Rev. M. J. Berkeley speaks of bread made from it, "better to my taste, and probably not less nutritious, than Cassava bread." These root-stems are also, on account of the quantity of tannin and astringent matter contained in them (which, by the way, would rather interfere with the asparagus flavour), much used abroad in preparing chamois and kid leathers. The alkalic properties of the fronds make them useful too in soap. Farther, the Bracken is not without a medicinal reputation: it is still retained in the *Materia Medica* as a remedy for worms, and a bed of the green plant is looked upon by country folk as "the sovereignest thing on earth" for rickets in children. Common as the plant is there do not seem to be many varieties.

VARIETIES.

crispa, — Arnside, J. M. Barnes.
multifida, — Levens, J. M. Barnes; Windermere, F. Clowes.
variegata, — Windermere, F. Clowes.

BRITTLE BLADDER FERN.

Cystopteris fragilis.* — BERNHARDI.

The Bladder **Ferns** (*kystos* is Greek for *bladder*) are so called because the indusium, **even in** age inflated or bulged out like **a hood, has** when **young** the look of a flask **or** bladder. The plant differs in this from the flatness (the sori **in both being** round) of the Poly-stichums and **Lastreas** with **which** it was formerly

---

* Polypodium **fragile (*Linnæus*),** Aspidium fragile, Athyrium fragile, Asplenium fragillimum.

ranked, under the general name of *Aspidiæ*. There
are three British species of Bladder Ferns : the Brittle
or Fragile — *Cystoperis fragilis*, the Alpine — *C. regia*,
and the Mountain — *C. montana*; but only the first is
really authenticated as belonging to the Lake Country,
no claim being made for *C. montana*, and the likelihood
of *C. regia* depending only on the following paragraph
in Moore's last edition :—" We have not seen a native
mountain specimen of *C. regia* unless it be one from
Saddleback (Blencathra), in Cumberland, gathered
many years since by Mr. S. F. Gray." There appears
indeed to be only one authenticated habitat of the
plant in England : that at Low Leyton, in Essex.

The Brittle Bladder Fern is of a very delicate and
grassy appearance, the root-stems spreading under
favourable circumstances into large patches of numerous crowns, each of which throws up a tuft of several
fronds, from six inches to sometimes a foot in height.
The stipes, erect, and rather more than a third of the
length of the frond, is brittle, dark, shining, with a few
small scales at the base. The fronds are lanceolate,
bipinnate; the pinnæ lanceolate; the pinnules ovate-
acute, cut more or less deeply on the margin, the lobes
furnished with a few pointed teeth. In some vigourous plants the pinnules are so very deeply cut as to
become pinnatifid, almost pinnate, the lobes themselves
then resembling the smaller pinnules nearer the apex
of the pinnæ and frond. The venation, from the delicacy of the frond, is very readily seen. In the ordi-
narily-sized pinnules there is a somewhat twisty midvein, giving off a side branch or vein to each of the

lobes into which the margin is cut, these veins again branching into two or more venules according to the size of **the** lobe, and each branch generally bearing a sorus at about midway of its length. **The** sori **are** thus generally numerous and rather irregularly placed, often becoming confluent and covering the whole under-surface ; but their number, and confluence, varying much, depending upon the various circumstances of growth. The sori are nearly circular ; the flask or bladder shaped (like a hood over them) indusia become **in** age torn or split at the point into narrow segments, turned back, jagged and fringe-like, the whole being pushed off by the ripening spores.

HABITATS. — Borrowdale (*Miss Wright* and *G. B. Wollaston*), Whitbarrow (*Wollaston*), Egremont (*Robson*), Ullswater (*Wollaston*), Arnside Knott (*H. D. Geldart*), Fairfield, Kendal, Windermere (*Clowes*), Kentmere (*Clowes*), &c.

VARIETIES.

angustata, — Whitbarrow, **F.** Clowes ; Arnside, J. Crossfield ; Sizergh, J. M. Barnes.

dentata, — Kentmere, F. Clowes ; Arnside, J. Crossfield ; Kendal Fell, **J. M.** Barnes.

interrupta, — Windermere, J. Huddart.

THE OBLONG WOODSIA.

Woodsia ilvensis. * — R. BROWN.

The genus *Woodsia* (so **called from** Mr. Joseph Woods) is the British representative of the group Peranemeæ. Of the two British species, the Oblong Woodsia and the Blunt-leaved or *alpina*, only the first is found in the Lake Country. Even that is very rare. It was first discovered in 1846, in a small quantity, on

one of the Westmorland mountains, by Mr. Huddart, who afterwards found some hundred plants near Scawfell, in Cumberland. The next year other stations were found in Westmorland **by Mr.** Huddart, both alone and in company with Mr. Clowes. In some of these places were only a few plants, but in one a great many and very fine.

The Oblong Woodsia is especially a mountain Fern, an annual, dying down to the ground in winter and coming up again in spring. Its caudex **is** short, erect

---

* Acrostichum ilvense (*Linnæus*), Polypodium ilvense, Aspidium rufidulum, Lastrea rufidula, &c.

or decumbent, furnished with a few scales on the crown, forming tufts, which in favourable circumstances grow into masses rather large in comparison with the diminutive nature of the plant. The stipes is scaly and articulated, or jointed, at a short distance from the base, so that in age the upper part with the fronds falls away, the lower part still adhering to the caudex. The fronds are seldom more than four inches high, oftener less ; their form is lanceolate, varying in breadth, pinnate, the pinnæ usually set on nearly or quite opposite in pairs, obtusely oblong, with the margin deeply lobed or pinnatifid. They are of a thick dull-looking texture, and are more or less clothed on both surfaces, but especially on the veins beneath, with minute bristle-like scales and shining jointed hairs, among which the sori lie almost concealed. The venation of the segments of the pinnæ consists of a rather indistinct midvein, from which the smaller veins, simple or branched, extend to the margin near which the sori are produced. The indusia are peculiar in that they are not placed as covers to the sori, but attached under them. When very young, indeed, they enclose them ; but later they split from above into narrow scale-like segments, not easily distinguished without a glass from the frond-hairs among which they lie. In the full-grown state the sori lie in tufts of hair-like scales formed of the torn margins of the indusium, the latter being attached to the frond at the point beneath the capsules. No other native Ferns possess a structure at all approaching to this.

THE FILM FERNS.

THE TUNBRIDGE FILM FERN.

Hymenophyllum Tunbridgense. — SMITH.

The Film Ferns — *Hymenophyllum*, so called from the two Greek words *hymen* — a film or membrane, and *phyllon* — a leaf, belong to the same group (TRICHOMANINEÆ) as the Bristle Ferns — *Trichomanes.* They are all small moss-like plants, the smallest of our native Ferns, distinguished from other (foregoing) Ferns by having their fructification on the margins of the fronds, and from each other by the form and nature of the involucres which surround the fructification. These involucres are deep urn-shaped pits, in which are contained the spore-cases, clustered around hair-like or bristly receptacles, which bristles are indeed the ends of the frond-veins projecting into the urns. In *Hymenophyllum* these bristles are always shorter than the urn ; while in *Trichomanes* (a British genus also, but not found in the Lake Country) they project more, so that the fronds become bristly when very full of spores. Hence the name of Bristle Fern. They are known also by the farther difference that the involucres of

Trichomanes are entire, and those of *Hymenophyllum* split lengthwise into two valves.

The Tunbridge **Film Fern** (named from its being first found **near** Tunbridge, in **Kent)** grows in matted tufts upon rocks in moist warm places, usually **carpeting** the damp surfaces of the rocks themselves, **but sometimes choosing the** mossy **ground, or living moss-like on the trunks of trees** — the black **wiry rhizomes or** creeping caudices interlacing themselves among their neighbour plants. The fronds are very **short,** from an **inch to three or** at most six inches long, membranous **and half-transparent,** almost erect, and of a **dull** dead-looking brownish-green even when at their freshest ; lanceolate or slightly ovate, pinnate, with pinnæ pinnatifid or bipinnatifid, and having their branches mostly on the upper side, though sometimes alternately on each side of the pinna. The fronds are virtually **a** branched series of rigid veins, winged throughout, ex-**cept** on the lower part **of the** short stipes, by **a narrow** membranous leafy margin. The sori are produced around the axis of a vein, which, as before said, is continued beyond the frond-margins, and enclosed in an urn-shaped indusium, involucre, or cover, consisting of two almost perfectly round (orbicular) compressed valves, spinosely serrate on the upper margin. It will grow well in pots in equal parts of peat and silver sand, scarcely caring for any other mould, but requires a glass, and constant but not stagnant moisture.

HABITATS. — Coniston, Buzzard Rough Crag near Wrynose (*Ray*), Hawl Ghyll near Wastwater (*Robson*), Ennerdale (*J. Dickinson*).

ONE-SIDED (OR WILSON'S) FILM FERN.

Hymenophyllum unilaterale. — WILLDENOW.

In *Hymenophyllum unilaterale* the pinnæ are what
is called decurrent in the upper part — that is to say,
they are prolonged beyond their points of insertion, as
if running downwards, so that the fronds appear to be
one-sided, **or** unilateral. The name of One-Sided
might therefore be employed to designate it just **as
well as if** not better **than** the cognomen of its dis-
coverer. Like *H. Tunbridgense*, it grows from numerous slender thready stems, into dense tufts, from
which spring a crowded mass of half-drooping brown,
green, or olive-coloured, semi-transparent fronds,
averaging from **three to** four inches in height. The
fronds are lanceolate and pinnate, the rachis is usually
somewhat curved, the pinnæ are one-sided, convex

above, and all turned one way, as already described, the outlines of the pinnæ wedge-shaped, digitately pinnatifid (like the fingers of a hand notched almost to the bone). The extreme or ultimate lobes are linear-obtuse with a spinulose-serrated margin. The fronds when luxuriant have a tendency to become branched. The veins are twice-branched, branching alternately from the rachis, forking again so as to extend a venule to each segment; and after leaving the midrib are furnished with a narrow membranous leafy wing or border (which the rib itself has not). The sori are collected round the free ends of the veins and contained in the urn-shaped covers or involucres, which differ from those of *H. Tunbridgense* in being more or less obviously stalked instead of sessile, and in having their valves entire instead of serrated at the upper margin. Mr. Clowes notices also a farther difference between the species: that the fronds of *H. Tunbridgense* are annual, "never *grow* more than one year;" while those of *H. unilaterale* are perennial, lasting for several years and annually renewing their growth, bearing spores year after year.

HABITATS. — Patterdale, Ambleside (*J. Bowerbank*), Stock Ghyll Force (*Miss Beever*), Dungeon Ghyll, Scaw Fell (Black Rocks and Great End), Bowfell, Ennerdale (*Dr. Dickinson*), Scale Force (*H. C. Watson*), Honister Crag (*Rev. G. Pinder*), Gatesgarth Dale, Borrowdale, Lodore (*Miss Wright*), Keswick, near Hawkshead (*Miss S. Cowburn*), Coniston Old Man (*Miss Beever*), Silverdale (*Miss Beever*), Dalegarth (*Robson*), High Stile (*Pinder*).

ROYAL, OR FLOWERING FERN.

Osmunda regalis. — LINNÆUS.

"At Loch Tyne dwelt the waterman, old Osmund. Fairest among maidens was the daughter of Osmund the waterman. Her light-brown hair and glowing cheek told of her Saxon origin, and her light steps bounded over the green turf like a young fawn in his native glades. Often, in the stillness of a summer's even, did the mother and her fair-haired child sit beside the lake, to watch the dripping and the flashing of the father's oars as he skimmed right merrily towards them over the deep-blue waters. Sounds, as of hasty steps, were heard one day, and presently a company of fugitives told with breathless haste that the cruel Danes were making way towards the ferry. Osmund heard them with fear. Suddenly the shouts of furious men came remotely on the ear. The fugitives rushed on. Osmund stood for a moment; then snatching up his oars he rowed his trembling wife and fair child to a small island covered with the great Osmund Royal, and helping them to land, bade them to lie down beneath the tall Ferns. Scarcely had the ferryman returned to his cottage, when a company of Danes rushed in ; but they hurt him not, for they knew that he could do them service.   During the day and night did Osmund row backwards and forwards across the river (or the lake ?), ferrying troops of those fierce men.   When the last company was put on shore, Osmund, kneeling beside the river's bank, returned heartfelt thanks to Heaven for the preservation of his wife and child. Often in after-years did Osmund speak of that day's peril ; and his fair child, grown up to womanhood, called the tall Fern by her father's name." So says the heart-thrilling legend, touching, in its conclusion, even to the scientific botanist, accounting for the name of the stateliest of our Ferns.   There is another supposition, however, that the name is derived from *os* and *mund*, Saxon words for *house* and *strength* or *peace*, though what house-strength or house-peace has to do with the Flowering Fern it is difficult to say. Why not even a third guess, hardly likely to be farther off than the others, that it has something to do with *Osmonds* — in old Saxon *iron ore*, for is it not found in the iron countries, in Durham and in South Wales, and in our own iron district of Cumberland, if not nearer than Egremont or Sea Scale, yet that is nearer than Loch Tyne and the river of the ferryman ?   Whatever the origin, however, of the name given it by Linnæus, the Royal Osmund is indeed the grandest of our Ferns, under all circumstances a handsome plant, but especially beautiful when, in very luxuriant growth, its fronds, loaded at their tips by the fertile panicles, are bent down gracefully until they almost reach the sur-

face of the water by **the** side of which they prefer to
grow. From these panicles, springing like clusters of
flowers from the ends **of** the fronds, comes its name **of**
Flowering Férn.

The fronds of **the** Flowering Fern grow to an aver-
age height of three or four feet, sometimes even to the
royal stature **of eight, ten,** or twelve, **and** six feet or
more across. The caudex **is** tufted, in very old and
vigourous plants forming **a trunk a** foot or more above
the ground, from the crown of which, whether it is
close to the ground **or** elevated, grow the fronds.
When young, these fronds **have** generally a reddish
stipes, with a bloomy surface, the bloom being lost at a
later period. They are annual, perishing before the
coming of winter, smooth and of a bright yellow green,
paler beneath, lanceolate in general outline **when**
mature, bipinnate, the pinnæ lanceolate or ovate-
lanceolate, with pinnules oblong-ovate, somewhat auricled at the base, especially on the posterior side,
bluntish **at the** apex, and finely serrated along the
margin. Some of the fronds are entirely barren, while
others have several **of the upper pinnæ transformed**
into terminal fertile panicles. Each short spike-like
branch of the panicles (or flower-clusters) represents one
of the pinnules, the spore-cases being collected on it in
little **knots,** more **or less evident, these** knots (or
nodules) corresponding to the fascicles of the veins.
This is very plainly seen **in** partially-transformed pinnules. The venation, as seen in the barren fronds,
consists of a stout midvein **giving** off nearly opposite
veins, which are forked once near their base, the venules

being parallel, slightly curved, and once or twice forked before reaching the margin, where they are lost. In the fertile parts of the frond only the midrib of the pinnules is fully developed, and the spore-cases are attached to a small portion of the veins, which becomes developed just to serve as a receptacle. The spore-cases are of a reddish-brown, nearly globe-shaped, shortly stalked, reticulated, and two-valved, the valves opening vertically.

The Flowering Fern grows naturally in wet, springy, or boggy places, not much above the sea-level in England, and sometimes on the sea-shore hardly above high-water mark. It is common throughout Europe ; is found in Asia, in Mingrelia, and in India ; in North and South Africa, in the Azores, and in Madagascar ; and in North and South America, in Canada and New-foundland, the United States, Mexico, and Brazil. It is of easy culture, needing only moisture and a peaty soil in any sheltered situation. The caudex is said to possess tonic and styptic properties ; according to Gerard, the " root " boiled or stamped, and taken with some kind of liquor, is " thought to be good for those that are wounded, dry-beaten, and bruized." In Cumberland and Westmorland, and that adjoining part of Lancashire which should be Westmorland also, it is known as the " bog-onion," and held in esteem as an external application for bruizes, sprains, &c. The caudices are beaten, covered with cold spring water, and allowed to macerate all night ; the thick starchy fluid thus formed being used to bathe the parts affected.

HABITATS. — Windermere (*T. G. Rylands*), Lough-

rigg and Skelwith, Colwith (*H. Fordham*), Ullock
Moss by Keswick, Whitbarrow, Scale Hill, Egremont,
Sea Scale and Gosforth (*Robson*), Irton, Millom,
Brantwood by Coniston, Yewdale under Coniston
Crags.

THE COMMON MOONWORT.

Botrychium Lunaria.* — SWARTZ.

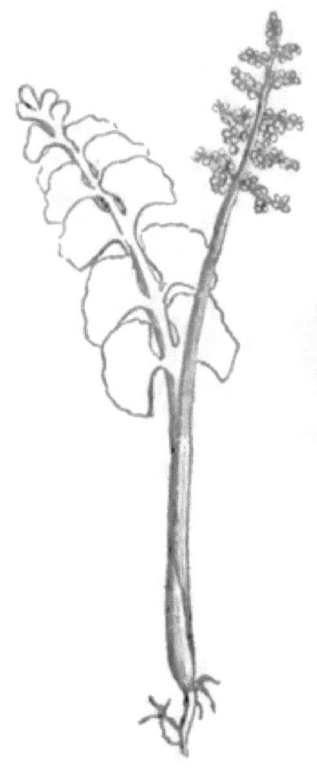

The Common Moonwort (a *Botrychium*, of the order OPHIO-GLOSSACEÆ, distinguished from all the Polypodiaceæ or True Ferns, by their young fronds being not circinate, but folded straightly, though at the same time resembling the Osmundineæ in having no elastic ring and in being two-valved) is one step farther in the course of natural variety, for, as through Polypodiaceæ and Tri-chomanineæ there is one regular progression and change of method of fructification from the spore-cases without indusium to the spore-cases with indusium, from the simplest forms of indusia to the flask or bladder shapes, from the spore-cases on the backs to the spore-cases on the margins, and the spore-cases (as in *Osmunda*) on the ends of fronds transformed into seeming stalks, so the Ophioglossaceæ show yet

* Osmunda Lunaria (*Linnæus*), Ophioglossum pennatum, &c.

one more change, the change into the appearance of a distinct flower-stalk being yet more marked, so much so as to be not at first sight distinguishable from the stalk of a veritable flowering plant.

Of the OPHIOGLOSSACEÆ there are two British genera — *Botrychium* or Moonwort, and *Ophioglossum* or Adder's Tongue, readily known from each other by their external features. Both genera have two-branched fronds, one branch looking like the leaf and the other like the flower; but they differ obviously in this, that *Botrychium* has its branches branched again, while those of *Ophioglossum* are simple and undivided.

The roots and caudex (or root-stem) of *Botrychium* differ essentially, says Mr. Newman, from those of the "True" Ferns. "The roots are stout, succulent, and brittle. The caudex is about the same size as the roots, perhaps rather stouter; it descends perpendicularly, and the roots issue from it at right angles. Before the plant has felt the influence of spring, the frond exists in a quiescent state, but perfectly formed. It then appears like a simple stem, scarcely an inch in length, and perfectly erect. On a closer inspection the component parts of the future frond will be clearly perceived; the stipes is swollen, and rather stouter than the upper part, the two branches of which face each other, the fertile branch of the frond being clasped by the barren or leafy part; and the fructification being thus entirely concealed, the uppermost pinnæ are incurved, as if to give still farther protection to the fruit. The whole is invested and completely enclosed in scale-like alternate sheaths, doubtless the decaying stalks of many previous years. As the spring advances, the frond rapidly increases in size, until in April it makes its appearance above ground, and in May or June attains its perfect development."* **Mr.** Newman also found the frond of **the ensuing year in every** respect **perfectly formed** — indeed, exactly **in the** state in which **it is found in** the early spring before development ; **while** the frond for the next following **year, though less** perfectly formed, also had the fruitful **and leafy** portions distinct **from** each other. These observations being made in May, while **the** plant was still growing, the fronds of three successive years were distinguishable **at the same time.**

The name *botrychium* is from **the Greek** *botrys* — **a** cluster, because of the likeness of the branched clusters of spore-cases to the form of a bunch or cluster of grapes. The English name of Moonwort **is given on** account of the lunate (or crescent-like) form of the **pinnæ in** the British **species.**

The Common **Moonwort** prefers dry, open, and elevated **pastures** and **waste** lands, and likes to skirt them under the shade **of** hedge-rows. It may easily be passed **over,** half **hidden as** it is **among** the herbage, for its **height** only **varies** from some two or three **inches to** six or eight ; but once **seen there is no** mistaking the double **row** of fan-shaped pinnæ which form its sterile branch. The lower half of **the** plant consists of a smooth, erect, cylindrical, **hollow** stipes, **whose** base is clothed by the brown membranous sheath which had covered it while **in** bud. Above are the two

* Newman, *History of British Ferns*, third edition.

separate branches of the frond,— one branch spreading, leafy, oblong, pinnate, with its crescent-shaped or fan-shaped pinnæ filled with a radiating series of twice or thrice-forked veins, one vein extending into each of the rounded teeth or lobes into which the margin is divided, —the other branch erect, fertile, compoundly branched, that is first divided into branches like the pinnæ, then again into further branches, on which, distinct, but clustered, the grape-like stalkless spore-cases are produced. The spore-cases are two-valved, and open transversely when ripe. The valves are concave. Occasionally, but very rarely, there are two branches: and a variety has the pinnæ pinnatifid.

The Moonwort is widely but sparsely scattered over the British Isles; and is found also in all quarters of the globe, including Tasmania and the Australian Alps in Victoria. It ranges from the sea-line to 3,000 feet above it. It has not generally been very successfully cultivated; but it seems may be, if taken up with a sufficiently large sod, and carefully kept cool and equally moistened. Even in the natural state it is unable to bear much drought.

HABITATS.—Keswick, near Aspatria (Rev. J. Dodd), Braystones, Muncaster Fells, Furness Fells above Coniston Water.

COMMON ADDER'S TONGUE.

Ophioglossum vulgatum. — LINNÆUS.

The genus Ophioglossum (Adder's Tongue, from the Greek *ophios* — **a** serpent or adder, and *glossa* — **a** tongue) is the type of the order Ophioglossaceæ, mentioned before as differing from the True Ferns in having ringless spore-cases and their spring fronds straightly folded.   It differs from *Botrychium* in having the branches of its two fronds quite simple **or** undivided, instead of being pinnate and bipinnate as **that is.   It** differs yet more markedly in that its fertile branches are not merely the branched panicles of *Botry-chium*, but distinct spikes in which the spore-cases are distichous (arranged in two rows opposite to each other), **like** the florets of many grapes.   **Yet a** third remarkable difference is noteworthy — that while in *Botrychium*, as already **seen,** the next year's fronds **are** found within the **bases** of the **growing** stems, in *Ophioglossum* a bud is developed **by the** side **of** this **year's** frond.

There are two British species of *Ophioglossum* — **the** Common **Adder's** Tongue — *O. vulgatum,*

and the **Dwarf**—*O. Lusitanicum*, the last only recently
found in Guernsey. The Common Adder's Tongue is
widely dispersed, and abundant where it occurs. **The**
only locality given for it in **the Lake** Country is in **the**
meadows by St. **Bees. It is scattered over the whole**
of Europe **and Asia, North America, and** Mexico, and
found in some of its varieties **at the Cape** of Good
Hope, in New Zealand, **and Australia.**

The **Common** Adder's Tongue is small and **stemless,
the stem only** represented **by** the central **crown of its
few** coarse brittle **roots.** The young **fronds, from six
to twelve inches high,** are produced in **May and** perish
by the end of the summer. The **stipes is** variable in
length, smooth, round, hollow, **and** succulent. **The**
upper **part is** divided into branches — **one branch** leafy,
entire, smooth, obtusely egg-shaped and slightly **variable**
in form, traversed by irregular-angled veins, **forming**
elongated meshes within which are smaller veinlets, —
the other branch **erect,** contracted for about half its
length, **forming a** linear slightly tapering spike, **in** the
substance of which, upon each **of its** two opposite **sides,
a line of** crowded spore-cases **is** imbedded. The **spore-**
cases **are** therefore considered **to** be produced **on the
margin of a** contracted frond. When **ripe,** the **margin**
splits **at** intervals corresponding with the centre of each
spore-case, so that **the** spike then resembles a double
row of gaping spherical cavities.

The leaves, **pounded in a** mortar, are said to yield "a
most excellent greene oyle, or rather a balsame for
greene wounds, comparable **to** oile of St. John's-Wort,
if it do not farre **surpasse it."** The plant prefers loamy

pastures and meadow land, where its abundance, dis-
liked by the cattle, by no means improves the grass.
It is readily cultivable, but likes the shade of surrounding herbage.

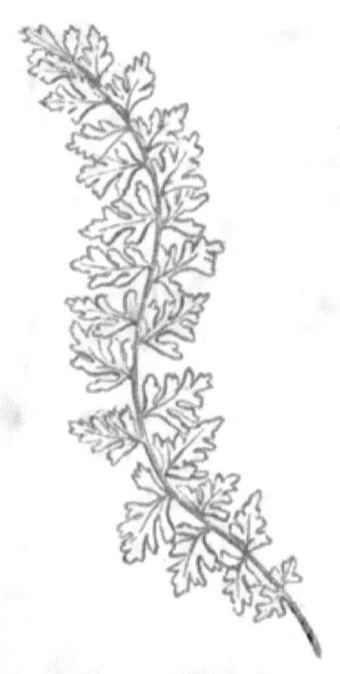

A. TRICHOMANES.

rar. incisum.

MEANINGS OF NAMES AND TERMS.

ALLOSORUS CRISPUS —

allos — *Greek*, differing ; *sorus* — *Latin*, from *soros* — *Greek*, a heap.

crispus — *Latin*, crisped, like parsley.

ASPIDIUM — L. — from *aspidion* — *G.*, a shield.

ASPLENIUM ADIANTUM-NIGRUM —

Asplenium — *L.* from *asplenion* — *G.*, spleenwort.

adiantum — *L.* from *adiantos* — *G.*, dry ; and *nigrum* — *L.*, black.

— *germanicum* — *L.*, alternate.

— *marinum* — *L.*, marine.

— *Ruta* — *L.*, rue ; *muraria* — *L.*, growing on walls.

— *septentrionale* — *L.*, northern (from the seven stars in the Great Bear).

— *Trichomanes* — *G.*, a hair or bristle ; and *manos*, soft, thin, porous, — or *mania*, excess.

— *viride* — *L.*, green.

ATHYRIUM FILIX-FŒMINA —

Athyrium — *L.* from *athyros* — *G.*, open (from the opening of the indusium).

Filix — *L.*, a fern ; *fœmina* — *L.*, feminine.

BLECHNUM SPICANT —

Blechnum — *L.*, *blechnon* — *G.*, a fern.

spicant — *L.*, spiked, growing to a point like a spike.

BOTRYCHIUM LUNARIA —

Botrychium — *L.*, *botrys* — *G.*, a bunch or cluster of grapes.

Lunaria — *L.*, lunar, lunate (referring to the crescent-like shape of the pinnæ).

CETERACH OFFICINARUM —

Ceterach, corruption of *Chetherak*, *Arabic* or *Persic* (meaning not known).

officinarum of *officina* — *L.*, officinal (used in medicine).

CYSTOPTERIS FRAGILIS —

kystos — *G.*, a bladder; *pteris* — *G.*, a fern.

fragilis — *L.*, fragile, brittle.

HYMENOPHYLLUM UNILATERALE —

hymen — *G.*, a membrane or film; *phyllon* — *G.*, a leaf.

unilaterale — *L.*, one-sided.

LASTREA ÆMULA —

Lastrea, the Latinized name of M. de Lastre.

æmula — *L.*, emulating, rivalling.

— *cristata* — *L.*, crested.

— *dilatata* — *L.*, dilated, spread out, broad.

— *Filix-mas* — *L.*, *filix*, fern; *mas*, male or masculine.

— *montana* — *L.*, mountain-growing.

— **remota** — *L.*, remote.

— *rigida* — *L.*, rigid.

— *spinulosa* — *L.*, spinulous, prickly-toothed.

— *Thelypteris* — *G.*, **thelys**, feminine; **pteris**, a fern.

OPHIOGLOSSUM VULGATUM —

ophios — *G.*, a serpent or adder; *glossa* — *G.*, a tongue.

vulgatum — *L.*, common.

OSMUNDA REGALIS —

Osmunda, derivation and meaning unknown.

regalis — *L.*, royal.

POLYPODIUM CALCAREUM —

polys — *G.*, many; *pous* — *G.*, a foot : many-footed.

calcareum — *L.*, chalky, limestone-growing.

— *Dryopteris* — *G.*, **drys**, an oak; *pteris*, a fern.

— *Phegopteris* — *G.*, *phegos*, a beech; **pteris**, a fern.

— *vulgare* — *L.*, vulgar, common.

POLYSTICHUM ACULEATUM —

     polys — *G.*, many ; *stichos* — *G.*, order (from the numerous orderly sori).

     aculeatum — *L.*, prickly.

     — *angulare* — *L.*, angular.

     — *lonchitis* — *L.*, a spleenwort.

PTERIS AQUILINA —

     Pteris — *G.*, a fern, from *pteron* — *G.*, a feather.

     aquilina — *L.*, eagle-like.

SCOLOPENDRIUM, from *Scolopendra*, the scientific name of a centipede.

WOODSIA ILVENSIS —

     Woodsia, Latinized from the name J. Woods.

     Ilvensis — *L.*, Elban, from the island Ilva, Elva or Elba.

---

acrogenous, — growing chiefly from the extremity.

acuminate, — extended into an acute terminal **angle.**

appressed, — pressed close, **lying** near the stem.

aristate, — bearded.

articulated, — jointed, separating readily at the joint.

auricle, — a small ear-like lobe.

caudate, — **with a tail ;** *ovate-caudate,* **an** egg with **a** tail, like a tadpole.

caudex (plural *caudices*), — the root-stalk or stem.

circinate, — rolled down, like a crozier-head.

confluent, — running into or uniting with one **another.**

cordate, — having lobes like the thick end of the heart in a pack of cards.

cotyledons, — the seed-lobes, the first leaves in the rudimentary plant or embryo.

crenate, — having the edges round-toothed.

crenulate, — with smaller teeth.

cruciform, — in the form of a cross.

deciduous, — **falling off, as the leaves of annuals.**

decumbent, — reclining upon the earth and rising again from it.

decurrent, — prolonged beyond the point of insertion, as if running downwards.

deflected (deflexed), — bent **downwards.**

deltoid, — triangular, like the Greek letter D — **delta.**

dentate, — **toothed.**

dicotyledonous, — having two cotyledons.

distichous, — **in two** rows.

dorsal, — placed upon the back.

dorsiferous, — bearing on the back.

echinate, — prickly, like a hedgehog.

endogenous, — **growing from within — increasing by** internal **growth.**

exogenous, — **growing** from without — by additions to the outer parts of the stem.

fascicle, — **a bundle, as of larch** leaves **growing from a common point.**

frond, — the combination of leaf and stem in ferns, &c.

glaucous, — bloom-covered, like a plum or **cabbage-leaf.**

herbaceous, — herb-like.

hippocrepiform, — horseshoe-shaped.

indusium (plural *indusia*), — the **membranous covering of the** spore-cases.

involucre, — a sort of **calyx or ring** inclosing an aggregate of flowers.

involucriform, — **divided at the** margin into hair-like incurved **segments.**

involute, — having the edges rolled in on each side.

lanceolate, — lance-shaped.

linear, — lying in lines ; **also** narrow, **with** parallel margins.

lobes, — the divisions or segments of a leaf; *lobules,* smaller lobes.

lunate, — crescent-shaped.

monocotyledonous, — having only one cotyledon.

mucronate, — abruptly terminating in **a** hard short spine.

mucronulate, — not so distinct a **spine.**

nodule, — **a** knot.

orbicular, — perfectly circular.

ovate, — egg-shaped; *obovate,* — inversely egg-shaped.

panicle, — a cluster of flowers.

paniculate, — having panicles.

peltate, — fixed to the stalk by the centre, or by some point distinctly within the margin.

persistent, — lasting, not deciduous.

petiole, — the stalk of a leaf.

pinna (plural *pinnæ*), — the leaflet or primary division of a pinnated leaf.

pinnate, — when simple leaflets (or pinnæ) are arranged on each side of a common leaf-stalk; *bipinnate,* when the pinnæ are again divided; *tripinnate,* a third division.

pinnatifid, — divided not quite to the stalk.

pinnule, — a small pinna — the secondary division of the leaf.

plicate, — folded lengthwise, like a lady's fan.

plumule, — the bud of a seed.

rachis (plural *rachides*), — the midrib or vein of a leaf or frond.

radicle, — the first root of a plant.

receptacle, — the part in which the organs of reproduction are placed.

reflexed, — curved very much backwards.

reniform, — kidney-shaped.

reticulated, — like net-work.

rhizome, — the creeping root-stalk.

rupestral, — growing in rocky places.

serrated, — with teeth like a saw.

sessile, — set on without any perceptible stalk.

soriferous, — bearing sori.

sorus (plural *sori*), — a cluster of spore-cases.

spicate, — in the form of a spike.

spinulose, — having spines or thorns.

spores, — the seeds of ferns.

stipes, — the proper stalk of the fern.

sub, — in composition means nearly, as *sub-ovate,* nearly egg-shaped; *sub-pinnate,* not altogether pinnate.

subnate, — under-growing.

suprasoriferous, — bearing **the sori on the** upper surface.

truncate, — terminating abruptly, as if **a** piece had been **cut off.**

tuberculate, — lumpy, wart-like.

venation, — the system of **veins.**

venules, — veinlets or small veins.

whole-coloured, — all of one colour.